# DYNAMIC
# TENSIONS

*Essays on*

Balancing Privacy, Security
& Identity in the 21st Century

# DR. TIM GODLOVE

---

# DYNAMIC TENSIONS:

*Essays on*

Balancing Privacy, Security
& Identity in the 21st Century

**GIANT STEPS PRESS**
Freeport NY 2013

Freeport, NY
2015
http://giantstepspress.blogspot.com

Godlove, Tim 1956-
Dynamic Tensions:
Balancing Privacy, Security
& Identity in the 21st Century

ISBN-13: 978-0692595602
ISBN-10: 0692595600

Printed in The United States of America.
Cover and book design by Norman Ball
gspressnow@gmail.com


First Edition
Giant Steps Press
Freeport, NY
All rights reserved

## Dedication

To Timothy, Jonathan and Laura, dynamic tensions in cyberspace who will inherit the world we create today, and who have seen the beginning and expect to see the end of the twenty-first century.

## Table of Contents

## Acknowledgements

This book is possible due to the leadership and inspiration of Major General Walter Stewart, U.S. Army (Retired), Major General James D. Davis, U.S. Army (Retired) and Captain Larry Meacham, U.S. Navy (Retired) with whom I served and from whom I learned.

I am grateful to Norm and Adrian as colleagues who made contributions to the book and for their wise counsel.

I especially wish to thank my family for the patience and understanding.

## Preface

The 21st century is an extraordinary time. The digital revolutionary transition is ahead and there are numerous ways we can steer future events. We need the wisdom to distinguish some technologies are blessings and others could undo the balance of privacy and information security.

This book is not for experts in the IT privacy, identity, security or mobility. The audience is ordinary people who are curious to know about the cyber world and the impact on the way we live our lives. People know to protect their wallets and checkbooks, as these provides a quick way for thieves to access money. People can be proactive in protecting files and electronic records.

I hope my knowledge, research and experience from this collection of essays provide you valuable insight. The dynamic tension of protecting data, including the unprecedented amount of your personal information, is the new challenge of the digital era. Gathering personal information and its power has only begun.

## Introduction

What is there left to *intrude into* if detection paradigms are already, *de facto*, everywhere at once? How many times can we be 'data-breached' until the vectors of our economic identity have all but been released to the public domain, at least among criminal elements? Frankly, are fears of intrusiveness even valid anymore if, at this stage, the digital world has managed to pervade every aspect of human activity? These are hard and troubling questions with no ready answers. Furthermore, they go to the existential core of who we are, both as public and private entities—and in the case of unwanted exposure, private entities drawn reluctantly into public view.

In IT, the intersection of privacy, security and identity presents a Gordian knot of overlapping priorities subject to continual negotiation between interested parties in government, industry and private life.

To be sure, recent developments in Privacy by Design (PbD) standards are duly welcomed efforts as they establish privacy as an integral component of IT system architecture. This proactive awareness of privacy as an embedded feature signals a 're-recognition' that human identity—the subjective self—must be safeguarded from uninvited inspection no matter what economic or security benefits are curtailed as a result.

Clearly, we can expect this uneasy truce to be continually contested by elements within our society, both lawful and criminal. Vigilance will be the key. At the center of this tug-of-war lies the issue of trust or what some parties have called 'trust tension'. Just as inadequate personal data creates authentication misgivings, excessive data gathering can intrudes on personal privacy.

Embedded herein too is a prickly 'eye of the beholder'

dilemma; one person's sense of invaded privacy (and privacy needs vary from person to person) is another's need to know for transactional integrity. There is no immutable standard. Dynamic tension is the way forward.

Fortunately, Dr. Tim Godlove provides us with some very thoughtful and eminently useful insights in the essays that follow.

Culled from nearly fifteen years of writing (the first appearing just a few short months after 911 when security in all things became a preeminent national priority, no less so than in the country's IT architecture), these essays span a multitude of topics with the majority addressing Dr. Godlove's primary career focus, the healthcare sector and Electronic Health Records (EHRs) where, as one can imagine, the tensions between privacy and security are especially acute.

Inadequate attention to compliance standards; a tendency to overlook more mundane and self-evident physical security issues in favor of encryption and technology-based measures; the pros and cons of Universal Patient Identifiers (UPI); the trade-offs inherent in the burgeoning Telework movement between widely dispersed data (and the security issues therein) and the obvious cost and quality-of-life advantages; the dystopian implications of biometrics versus their powerful authentication advantages; hopefully, this very cursory list whets the reader's appetite as to the breadth and topicality of the IT issues addressed in these pages.

Finally, this is welcome collection to any IT Manager's library. We can only hope Godlove continues to offer his cogent analyses of prevailing industry trends in the challenging years ahead.

*—Norman Ball, MBA, PMP*

## My Initial Exposure to the Industry

October 2015

I was introduced to mainframe computing and computer programming as a computer operator in the late 70's working for National Aeronautics and Space Administration at Goddard Space Center. I was processing Landsat and Nimbus 7 satellites data on the weekends. The work included access to computers, networks, and stored information to process images focused on giving scientists the ability to assess changes to the Earth. Clearly, computers were going to play a role in saving the world. The digital era was beginning and with every new technology, there can be misused. Modern technology and digitalization are not what is seems. The unseen and silent exchanges of medical, financial, genetic, and employment records are increasingly being passed back and forth over information systems machines without us knowing.

The digital age signifies excitement and uncertainty, potential and risk, and threat and opportunity to the courageous who venture out onto the digital superhighway. All of us are so to speak, fellow explorers, and we should be curious about the condition of our vehicle, the skills of its designers and operators, and about any impediments we may encounter. The functions of human enterprise rapidly are being digitized, interconnected over networks, and stored and processed within information systems. This is why cybersecurity matters and there are implications of cybersecurity in a world of growing cyber threats. The term "personal information" has become somewhat of an oxymoron in the modern digital age, in that very little information is personal anymore. Data such as medical, financial, and employment records can be accessed by technologically adept people even when such files are not open for public inspection and are protected by laws and regulations.

In addition to the lack of privacy that accompanies modern technology, technological solutions are becoming particularly useful in making it more and more difficult for a person to conceal their identity. What is often raised is "the enduring conundrum over who can be trusted in cyberspace," or in any digital transactions. For that matter, is being exacerbated by technologies that unearth concealed identities. Weak forms of digital identities are already broadly used in the form of bank account and social security numbers. They provide only limited protection, for it is a simple matter to match them with the individual they represent. As such, concerns about hacking, theft, cyberterrorism, privacy, identity and anonymity in the modern digital era are becoming increasingly analyzed, discussed and speculated upon.

Ironically, it is not only "bad guys" that are causing concerns about the obliteration of life as we know it, but the expanding surveillance ability of government and law enforcement agencies is caused for apprehension as well. Surveillance and biometric technologies have advanced exponentially in recent years, making it difficult to avoid seeing images of an Orwellian society looming closer than ever before. While there are many benefits to these technologies, there are also many concerns.

Cybercrime and forensic evidence related to biometrics are becoming an increasingly popular method of identifying unique human characteristics as a means of authenticating an individual's identity. Where it used to be employed exclusively in crime-solving endeavors and high powered corporate or government security, the science of biometrics is quickly becoming about as commonplace as the personal computer.

The science of biometrics is ultimately based upon the analysis of distinctive physical traits, such as fingerprints and retinal scans; as well as personal characteristics such as physical, biological and behavioral patterns. Examples of personal characteristics include voice pattern recognition and handwriting analysis. The overall goal of biometrics is to use modern technology to identify individuals, and authenticate their identity, in a more effective and efficient manner. Permitting biometric recognition can be used in the identification mode or the verification mode. In the identification mode, the system identifies a person from the entire population by searching a database for a match. In the verification mode, the biometric system authenticates a person's claimed identity from his or her previously enrolled pattern.

In the business world, biometrics is primarily used as a security measure to prevent unauthorized personnel from accessing confidential data. In business, biometrics is based "not on what the user knows, or what they carry, but who the user is, some unique characteristic. One method that is becoming more and more popular is keystroke analysis, which "authenticates the user based upon their typing characteristics."

Keystroke analysis studies keystroke latency or the time between successive keystrokes, as well as hold-time characteristic, or the time to press and release a key. These factors are unique to individuals, especially as research has led from pattern recognition approach such as linear and non-linear distance techniques, z-tests and Bayesian classifiers, to algorithms such as the feed forward multi-layered perception's algorithm, the radial basis function algorithm, and the generalized regression neural network among others. Using neural network classifiers to perform classification with an error rate of only about 12%, suggesting that this approach provides cell phone users with more and better security. Overall, the investigation has shown that ability for classification algorithms to be correctly discriminate between the majority of users with a relatively good degree of accuracy based on the hold-time of a key.

The analysis goes on to describe how the data collection, classification and authentication engines would work without inconveniencing the user. The system is best used by users who use cell phones regularly and is not for users with "large variations in their handset interactions." In the future, cell phones with built in videoconferencing cameras could adapt facial recognition to strengthen mobile security.

Biometrics is not just used throughout the business world, but it is becoming a part of home security as well. For example, the Biometric security systems, like the fingerprint scanner available on the IBM ThinkPad T43, is becoming more common for home use. Furthermore, biometric door locks are becoming increasingly popular in home use because they are convenient (no need to fumble around looking for keys), and they are more reliable security-wise than traditional locks. They can be used by scanning one's fingerprints, retina or other parts of the body that make an individual entirely unique.

Even public schools are not immune to the growing biometrics trend, as the scanning of the literal 'student body' is becoming commonplace. Some schools use portable scanners to collect digital images of the students' fingerprints, which need to be regularly updated as the students grow and their fingers change. Biometrics is used for everything from the authentication of new transfer students to providing the ability to buy lunch in the cafeteria without cash, to checking out books from the library to recording student attendance.

Biometric technologies have advanced exponentially in recent years. While many people are concerned about privacy issues, the technology is not slowing down because of these concerns. Thanks to modern technology we can identify an individual based on his voice, his speech patterns, the way he walks and stands, and the patterns in the retina in his eye. Scientists thought it was a huge breakthrough when computer technology made it possible to compare fingerprints taken from the scene of a crime instantly with an entire database of stored fingerprint files – and it was. Even so, now that just seems like "old news" because biometric identification has come so incredibly far in recent years.

There are two primary ways that biometric recognition

technology can be utilized. The first is regarding identification, in which "the system identifies a person from the entire population by searching a database for a match." The second is what is known as the verification mode. This is a faster mode of matching personal traits to individuals because it only looks at patterns that have already been entered into the system. Whereas it used to be that an officer had to rely on lifting fingerprints from the scene of the crime, there are so many new ways technology can identify criminals whom the ability to match clues with perpetrators has become an almost instantaneous possibility. As such, it has revolutionized crime solving, including cybercrimes.

The Internet has made access to information remarkably easier for billions of people. Unfortunately, with this ease of access also comes a greater likelihood of cybercrime. One incident of cybercrime that recently made headlines was the so-called "Comcast Hacker Case." In this incident, two young men named Christopher Allen Lewis (age 20) and Michael Paul Nebel (age 28) along with co-defendant James Robert Black, Jr., hacked into the website of Comcast at www.comcast.net and redirected traffic to their own website. The incident took place in May of 2008, but the sentencing just occurred in September 2010 ("Comcast Hackers," 2010).

The defendants went by cyber nicknames, which is a common occurrence in hacker situations. Christopher Allen Lewis of Newark, Delaware called himself "EBK." Michael Paul Nebel of Kalamazoo, Michigan called himself "slacker," and their co-conspirator James Robert Black, Jr. went by "defiant." Collectively, this group of hackers had labeled themselves Kryogeniks ("Comcast Hackers," 2010).

In directing over 5 million Comcast customers to their Kryogeniks website, customers were unable to read or listen to their mail. Instead, they were greeted with the message "KRYOGENIKS Defiant and EBB RoXed COMCAST sHouTz to VIRUS Warlock elul21 coll1er seven." The cost to Comcast for this stunt/crime was approximately $90,000.00. Not surprisingly, the FBI got involved in the matter, and criminal charges were filed ("Comcast Hackers," 2010).

Lewis and Nevel, both of whom pled guilty to conspire to disrupt service at Comcast's website, were sentenced in on September 24, 2010, to 18 months in prison. Also, they were ordered to pay Comcast back for the estimated $89,578.13 that they lost due to this crime. Their co-conspirator, Black was only sentenced to four months in prison after his case was transferred to the Western District of Washington, but he was also ordered to contribute to the restitution payments ("Comcast Hackers," 2010).

The way these hackers pulled off their crime was that they gained control of the domain by phone and sent a hacked e-mail from a Comcast account to Network Solutions (Comcast's domain registrar). This allowed them to gain control of 200 Comcast domains. In an interview the day after the attack, Defiant and EBK told Wired magazine's Threat Level that they didn't initially set out to redirect the site's traffic. Instead, they merely changed the contact information for the Comcast.net domain to Defiant's e-mail address; for the street address, they used a false address using crass language. Whether they intended to redirect traffic and shut down Comcast for 5 hours or not, the result was devastating to Comcast nonetheless.

So what was the motive behind this cybercrime? Apparently, it was a "ego" crime, which expanded into what some would call a "revenge" crime. The men called the Comcast manager at home to brag about what they had accomplished. However, the manager did not give them the reaction they had hoped for. Instead he "scoffed at their claim" and ended up hanging up on them. Angry and insulted by this reaction, the men decided "to take the more drastic measure of redirecting the site's traffic to servers under the hackers' control."

There has been some speculation that the attack was motivated by Comcast's reported sabotage of BitTorrent traffic. However, the defendants denied that this was the case. In fact, their explanation for the attack was that "they just hate Comcast in general. Defiant sarcastically quipped, "I'm sure they hate us too."

This seems to be a case of some mischievous young men getting in the way over their heads and not realizing how serious the consequences of their actions would be. Early in the investigation, they were apparently laughing about the incident and enjoying their stint in the media spotlight. Now that they have been sentenced to serious jail time and are forced to pay major restitution, the seriousness of their crime has undoubtedly begun to sink in.

While there was no new legislation specifically attached to the Comcast incident, crimes such as this demonstrate that computer security is an enormously difficult problem for which no simple solution exists. Obviously, there are differences between detecting an intrusion attack and preventing one from occurring in the first place. Preventative measures are certainly more helpful and less complicated. However, with new intrusion techniques cropping up all the time it is virtually impossible to predict precisely what needs to be secured and in what fashion.

Some people claim that the government needs to step in to do more to combat computer viruses, because not only can they halt the operations of businesses like Comcast, which slows the economy, but they could potentially wreak havoc on the government itself. The 2007 movie "Live Free or Die Hard" illustrated just how vulnerable our entire society is an environment ran almost exclusively by electronics. Anyone with a great deal of computer savvy and the wrong intentions could potentially bring our world to halt. The Comcast incident is merely one of the many symptoms of the potential for mass destruction that professional hackers are capable of invoking.

While the scenario portrayed in Die Hard 4 were somewhat extreme, computer security breaches happen every day, ranging from annoying spam email directed at an individual, to intrusion attacks designed to infiltrate computer networks to gain access to sensitive and confidential data. The number of attacks against corporations, as well as the government, continues to increase despite the plethora of measures continuously devised to combat the intrusion.

Many people tend to think of the corporate giants like Comcast as being somewhat immune to routine security problems, simply because they have the financial resources at hand to combat them. However, it is becoming increasingly clear that money cannot buy security any more than it can buy happiness if the impenetrable device does not yet exist.

It seems with every innovation in security comes an innovation in security breaching, and the race on both sides of the fence seems infinite. Security measures and forensic techniques are becoming more advanced but have yet to surpass the rate at which intrusion technology is growing. When three young men just fooling around can cost a company $90,000 in five hours, it seems clear that the hackers are staying far ahead of the corporations when it comes technological know-how.

All in all, at present, readily available network security components such as Firewalls, Anti-Virus programs and Intrusion Detection Systems (IDS's) are unable to combat the vast range of malicious intrusion attacks perpetrated on computer networks and systems effectively. This is where the need for proactive security measures comes into play and becomes a necessary part of making and keeping information technology secure.

As the biometrics trend continues to grow, there is some concern about privacy issues, which are especially sensitive when minors are involved. For example, reports in the article Learning to Live with Biometrics, that Chris Hoofnagle, associate director of the Electronic Privacy Information Center in Washington, D.C. believes that fingerprint scanning schools "sets a dark precedent, conditioning students at a young age to embrace the idea of Big Brother-style biometric tracking...If ever there was a generation that would not oppose a government system for universal ID, it's this one.

It is certainly understandable that biometrics would conjure up images of futuristic Orwellian disaster scenarios. However, the main complaint of using fingerprint scanning schools is not about privacy but a lack of efficiency and convenience. Mr. Bob Engen, president of Educational Biometric Technology, suggests that "speed, not security or privacy, seems to be students' biggest concern with the system. The fingerprint-recognition systems tend to run slowly - slower than manually punching in a number, for instance - if a school is using a computer that is more than a few years old. Additionally, large student populations can slow the system since it has to run through every stored image before identifying the best match."

Clearly, there are still adjustments to be made for biometrics technology is used pervasively in businesses, homes and schools. However, modern biometrics are machine readable and recordable. Biometric information recorded by machine, and the data linked to biometric observations, can be copied easily, shared quickly and widely, combined, and stored for long periods of time without degrading. Biometrically authenticating identification by machine is highly accurate and, more important. It is highly usable personal information. It allows institutions to collect data and index it to precise and highly accurate human identifiers, making it useful in countless ways.

Currently, businesses are the most prevalent users of biometric technology as compared to homes and schools. However, the popularity of this technology in all realms is continuing to escalate. Does this mean we are headed for a "Big Brother" type of society? Or are we merely looking forward to a more secure, more convenient and more efficient society? Only time will tell; however, if the present is any indication of the future, modern technology will soon advance beyond the point where virtually no human function or interaction will be technology-free.

Identity concealment when technologies designed to conceal individual identities are advancing, yet they are still not the point that they provide full protection. The advent of smart cards that generate changing pseudo-identities will facilitate genuine transactional anonymity. 'Blinding' or blind and digital signatures will significantly enhance the protection of privacy. A digital signature is a unique 'key' that provides, if anything, stronger authentication than my written signature. A public-key system involves two keys, one public, the other private. The advantage of a public-key system is that if you are able to decrypt the message, you know that it could only have been created by the sender. Modern technology offers a variety of ways of uncoupling verification from unique identity. Validity, authenticity, and eligibility can be determined without having to know a person's name or location. Public policy debates will increasingly focus on when verification with anonymity is, or is not appropriate and on various intermediary mechanisms that offer pseudonymous buffers but not full identification.

Those who are concerned that privacy protection must be strengthened a call for government action to require organizations that have confidential information to encode all data and impose electronic barriers against intrusion. This could require government supervision and registry of coding keys. Those who believe that government regulation is not needed to argue that voluntary measures by such organizations will be sufficient and point out that, to a great extent, privacy is already compromised since personal information is shared by insurance companies and credit-card companies whenever a person uses medical insurance or applies for credit. In these cases, people waive their rights to privacy in order to obtain benefits. There is a concern that imposing government interference could lead to the danger of a police state.

Threats of terrorism have, of course, exacerbated the quest for technology that makes identity concealment impossible. This logical considering that the 9-11 terrorists succeeded in large part because they could slip through the system. While the concept of national ID cards emerged long before September 11, 2001, the 9-11 tragedy incited a strong resurgence of demand and controversy.

The ACLU (2015) cautions that if a national ID card and database were introduced, "the linkage of government databases with corporate databases increases the likelihood that intimate personal information, credit histories, spending habits, unlisted telephone numbers, voting, medical and employment histories-could be easily accessed without a person's knowledge." The ACLU characterizes ID cards as a blatant invasion of privacy and suggests that a computer registry represents a direct violation of Americans' basic civil liberties.

The main concern about the introduction of national ID cards centers on the ongoing debate between the rights of privacy and the security of the people. However, apprehensions also exist regarding elitism. According to the Anarchist Federation, "ID is a class issue – the rich will ensure their anonymity by their limited need for the welfare state. Most recently we have heard that children of 'celebrities' (which will undoubtedly include well-known politicians!) will be exempted from the Children Index - yet another clear message that ID will not affect everyone in the same way. As well as money, power and influence will give the upper classes anonymity from the state and the private companies which will run identity databases."

Few things are of more importance to Americans than their rights. The right to privacy has always been held sacred by Americans, but at the same time, the right to have access to knowledge is valued as well. As such, there is a perpetual conundrum relating to identity protection and anonymity in the sense that it is impossible to have it both ways. Protection of privacy and the free distribution of information are innately contradictory ideals. People want the right to know what the government is doing at all times, but the idea of the government knowing what they are doing at all times seems appalling. People are assured that they conduct transactions safely and securely in cyberspace, and yet the amount of information that Internet companies are able to access about their users is phenomenal. Essentially we are living in an age of constant contradictions between what we want technology to do for us, and what we fear that it will do to us. Another prime example is cyberterrorism.

"Cyberterrorism" is the ability to access, manipulate and/or destroy vital government data. There is, however, debate over the level of importance, which should be attributed to the U.S. government's efforts to plan successful defense strategies against cyberterrorism. This debate stems primarily from the level of seriousness with which threats of cyberterrorism are afforded. As Anthony Cordesman author of Cyber-Threats, Information Warfare, and Critical Infrastructure Protection: Defending the U.S. Homeland in 2002:

"U.S. military and defense officials involved in information warfare and planning and executing cyber-war have divided views. Those directly involved in cyber-offense, however, generally seem to feel that carrying out a successful major cyber-attack is far more difficult than those outside the national security arena recognize. They do not minimize the risk of cyber-attacks, but they feel they will have limited impact and that many if not most critical systems are isolated, difficult to identify and enter in concerted attacks, and can be reconstituted within an acceptable time frame and cost. This disconnect between defense and offense illustrates a basic problem underlying both any unclassified analysis of cyber-threats and their impact on homeland defense and a critical gap in the federal response to these threats."

Just how critical threats of cyberterrorism are, however, continue to be a subject of debate. Part of this debate has arisen from a lack of clarity as to exactly what cyberterrorism entails. Cyberterrorism is considered to be unlawful attacks or threats against computers and networks in order to further some political or social objective. This is a rather vague perspective that needs to be redefined to include any use of technology to undermine the safety and security of our nation. After all, the vulnerability of the United States to cyberattacks suggests that the next significant terrorist event in this country may be coupled with some form of cyberterrorism. Therefore, it is important to understand what cyberterrorism refers to, what such acts might involve and what the United States is doing to lessen the impact such as attacks might produce.

The world has certainly experienced more change in its economic, cultural, social, scientific, and political systems in the last 40 years than at any time in history. Moreover, the rate of global change due to satellites, telecommunications, information technology, electronic fund's exchange, and other forms of advanced science and technology will, if anything, continue to accelerate ever more rapidly in the years ahead. In short, this is not an easy time for a world shrunk by technology and yet fractured by fundamental religious beliefs and terrorism.

The tragic attacks on the World Trade Center and the Pentagon on September 11, 2001, have made it all too clear that wars do not have to be declared, and threats do not have to be overt. It is brutally clear there is a wide spectrum of potential threats to the U.S. homeland that do not involve the threat of overt attacks by states using long-range missiles or conventional military forces. Such threats can range from the acts of individual extremists to state-sponsored asymmetric warfare. They can include covert attacks by state actors, state use of proxies, and independent terrorist groups. They can include attacks by foreign individuals and residents of the United States whose motives can range from religion to efforts at extortion. Motives can range from well-defined political and strategic goals to religion and political ideology, crime and sabotage, or acts by the psychologically disturbed. The means of attack can vary from token uses of explosives, cyberterrorism, car and truck bombs, and passenger airliners to the use of weapons of mass destruction.

While these threats are of primary concern, the focus of this paper, cyberterrorism, seems to be causing the most controversy regarding its criticalness. Part of this debate stems from a lack of clarity as to exactly what cyberterrorism entails.

Edward Snowden is probably the most famous (or infamous depending upon your view) "cyberterrorist" of all time. His release of hundreds of confidential documents about the U.S. government's privacy invasions and surveillance techniques resulted in him fleeing to Russia, where he remains a fugitive. In addition to the hero/villain debate is the confusion over whether Edward Snowden qualifies as a cyberterrorist. There are several different definitions of a cyberterrorist that can be examined in relation to Snowden. One definition is that cyberterrorism is concerted, sophisticated attacks on networks. This definition as "very broad and all-inclusive." Another definition of cyberterrorism that incorporates the type of motivation, the purpose (or desired outcome) of the attack, and the objects of the attack. Since motives and outcomes are a huge part of the Snowden debate, this definition should prove quite valuable in settling the question of whether Snowden is a cyberterrorist. The definition in the digital era is the convergence of terrorism and cyberspace. It is understood to mean unlawful attacks and threats of attacks against computers, networks, and the information stored therein when done to intimidate or coerce a government or its people in furtherance of political and social objectives."

The first part of this definition is problematic in attempts to attribute cyberterrorism to Snowden because it involves "threats of attacks against computers, networks, and the information stored therein." Snowden's actions did not attack any of these things. He did not hack into the computer and steal the information; he was granted access to it. He worked among these people. He just was not supposed to tell anyone what he knew. So, essentially he broke a promise. It was indeed an extremely serious promise. Even so, if breaking a promise was equal to terrorism, then every president who was ever in office would be considered a terrorist.

Regarding the second part of the definition, the motivational part, Snowden does not fit into that understanding of cyberterrorism either, unless one believes that Snowden's actions were self-serving, which has been demonstrated above that this does not seem to be the case. Snowden did not release the documents "to intimidate or coerce a government or its people in furtherance of political and social objectives" he did it to stop the government from further its own political and social objectives while trouncing on the rights to privacy that the people hold so dear.

Ironically, it is because the American people value privacy so intently that so many Americans are outraged by what Snowden has done. By exposing the government's lack of regard for privacy rights, he has simultaneously circumvented the government's right to privacy by exposing its secrets. So instead of the people being angry with the government for stepping beyond the boundaries of their surveillance rights, they are angry with the whistleblower who, for all intents and purposes, beat the government at its own game. As the laws and norms are designed to protect the privacy of individuals, often run counter to recommendations for widespread monitoring of public facilities. The misuse of governmental power in search of public security is itself a threat. Finding the appropriate balance between the two goals in a democratic society will likely require continued adaptation as conditions of threat and administrative performance vary.

Cyberterrorism is considered to be unlawful attacks or threats against computers and networks in order to further some political or social objective. This is a rather vague perspective that needs to be redefined to include any use of technology to undermine the safety and security of our nation. After all, the vulnerability of the United States to cyberattacks suggests that the next significant terrorist event for this country may be coupled to some form of cyberterrorism. Therefore, it is important to understand what cyberterrorism refers to, what such acts might involve and what the United States is doing to lessen the impact such as attacks might produce. These goals constitute the primary purpose of this paper.

Long before the infamous World Trade Center attacks, cyber-crime and cyberterrorism were major problems for both governments and businesses. These problems can be neatly summarized (at least in the business context) by a full-page advertisement placed by IBM in the October 1996 edition of The Atlantic Monthly: "Will a 14-year-old Sociopath Bring My Company to Its Knees?" The question is a good one. Financial institutions in the City of London paid ransom in 1996 to a gang of cyberterrorists who threatened to wipe out computer systems (Sunday Times Insight Team, 1996). The gang received $777 million, including money from banks, broking firms, and investment houses in the United States. According to the United States National Security Agency (NSA), the terrorists used advanced information warfare techniques such as electromagnetic pulses, high emission radio frequency guns, and "logic bombs," remotely detonated coded devices.

Clearly the prospects of cyberterrorism even ten years ago were very real and alarming. Another pertinent example involves Eli Biham of Technion, the Israel Institute of Technology in Haifa, and Adi Shamir of the Weizmann Institute in Rehovot, who discovered a code-breaking technique that can crack virtually any encryption system. The technique is called differential fault analysis (DFA) and can be used to break encryption systems, such as the Data Encryption Standard (DES), which all banks use for commercial transnational communications and transactions.

Furthermore, in September 1996 a hacker closed down a thousand web sites on the Internet, in an attack aimed at the provider Panix. The attack involved firing hundreds of requests per second for information to nonexistent addresses. As each request was treated as "innocent until proven guilty," the computer "clogged up." William Cheswick of Bell Labs in Murray Hill, New Jersey, believes that there is no satisfactory way at present of dealing with this difficulty given the way the Internet now is organized. It will have to be dealt with by the adoption of a new version of the Internet's protocol. According to the United States established squad of counter- cyberterrorism, Cyber Security Assurance Group, hackers had entered unclassified Pentagon computer systems 250,000 times in 1995 and were successful 162,500 times (Casey, 1996). Jim Settle, former head of the FBI's computer security section, believes that a dedicated band of cyberterrorists could bring the United States to its knees.

Steve Orlowski, Special Advisor in Information Technology at the Australian Attorney-General's Department, has said, "it is now becoming recognized that the threat cyber warfare poses to non-military infrastructures is as great, if not greater than the traditional military threat." Orlowski goes on to say in his interview that, "one nation can obtain the advantage over another by destroying confidence in either the technology itself, thus reducing the rate of development of that technology and any economic advantages of it, or the uptake of the technology, often the subject of massive investment, with similar consequences."

Information technology, which was supposed to be able to supply us with a secure exchange of information, is generating its own critical instabilities. G. J. E. Rawlins, in his book Moths to the Flame: The Seductions of Computer Technology, observes that Murphy's Law operates with respect to computer technology because "our largest systems are too complex for us to completely predict their behavior. We've lost control. Rawlins goes on to write:

In a complex system too many things can interact, and so too many things can go wrong. No programmer can predict them all. Today's computers can't help us either, because they don't understand what we want--because we don't understand what we want. Nor can they tell us we're bungling a program because--so far—they far—they far—they far--they can only do what we tell them to do, and we don't see the error. If we did, we would simply fix it.

These problems have always existed, but they have increased in their intensity since the infamous 9-11 attacks in 2001. Clearly, there is an ongoing need to develop plans of counterattack both now and in the future in order to curtail this escalating and frightening trend. Electronic combat can be considered within the same intellectual framework that rules the relevant geographies for land, sea, air, and space warriors. Of course, cyberspace is different, and cyber power can directly wreak damage only in cyberspace. However, technically, the general strategies of warfare apply no less to cyberspace than they do to the 'real world'. Therefore, the first step in combating cyberterrorism is to understand exactly what we are dealing with.

The number of attacks against the government and corporations continues to increase despite the plethora of measures continuously devised to combat the intrusion. Security measures and forensic techniques are becoming more advanced but have yet to surpass the rate at which intrusion technology is growing. At the present time, readily available network security components such as Firewalls, Anti-Virus programs and Intrusion Detection Systems (IDS's) are unable to combat effectively the vast range of malicious intrusion attacks perpetrated on computer networks and systems. This is where the need for more advanced security measures comes into play.

Security measures for proper and effective network security provides the following:

• Accountability-proof that an intended transaction indeed took place.
• Confidentiality-protection of confidential information from an eavesdropper.
• Integrity-assurance that the information sent is the same as the information received.
• Authority-assurance that those who request data or information are authorized to do so.
• Authenticity-assurance that each party is who they say they are.

The simplest forms of security are password control and firewall protection. The best type of password is one that incorporates letters, numbers, and punctuation; allowing simple or predictable passwords (such as birthdays) create a security risk. A firewall is software or hardware that limits certain kinds of access to a computer from a network. By forcing all traffic to and from the

Internet to flow through a firewall, risks to the local network are decreased. Nevertheless, malicious scripts and applets can be used to infiltrate firewalls, intercept passwords, and wreak havoc. Blocking software can be used to prevent viruses from entering a system in the first place when prevention measures come "too late." However, computer forensics can be a valuable tool for "solving the crime."

Cyber forensic is the discovery, analysis, and reconstruction of evidence extracted from any element of computer systems, computer networks, computer media, and computer peripherals that allow investigators to solve the crime. In essence, it follows very similar procedures as those followed by traditional forensics experts when attempting to solve crimes. Also, cyber forensics focus on real-time, on-line evidence gathering rather than the traditional off-line computer disk forensic technology.

Analogous to conventional criminology forensics, contemporary computer forensic is generally performed after the incident in question has already occurred. Information obtained from computers has been used for the U.S. court system for decades. However, while at first, the courts treated the evidence, just as they had always treated traditional forms of evidence, as technological advancements continued to expand, it became increasingly clear that the rules must be written to compensate for the whole new set of circumstances that cyber technology has invoked.

In a distributed networked, environment it is crucial to carry out forensic examinations and evaluations of the victim's information systems regularly so that the damage can be thwarted before occurring, instead of just managing the damage afterwards. This is critical for the secure and ongoing operation of important information systems and infrastructures.

Unless special intrusion detection technology is installed, in most cases of intrusion attacks, "the identity, location and objective of the perpetrator are not immediately apparent." The attacks that are being made can come from a variety of sources, and while the reason behind the attack is irrelevant in regard to the end result, such as information can provide significant details for protection against future attacks.

Decision made yesterday and decision made now related to digitalization could haunt us in the future. Effective security involves more than obtaining the right technology. The real challenge is assuring its effectiveness, surrounding it with policies and practices that reduce the risk, and addressing a changing environment. Traditional approaches to information security offer little help considering that the basic principles of encryption, authentication, and security that provide the foundation of most Internet security systems are available in a variety of ever-evolving forms. Good security must be a dynamic process that addresses a constantly changing environment. This requires a steady flow of information and analysis around emerging security issues, to protect against new threats before it is too late. Even risk transfer mechanisms must regularly be examined to ensure that new threats are covered. Policies, practices, configurations must be updated dynamically to remain relevant.

In the digital age information technology and the cyber world provide an ideal base for a technological revolution having a major impact on the way we live our lives and what our lives are.

Concerns of Identity and Anonymity in a Digital Era
The Security Journal, Fall 2010

*A right of anonymity is a troubling legal concept, especially when it comes to information that may cause damage, and the person or institution originating the message may need to be held responsible.*
—Anne Wells Branscomb

## Introduction

The term "personal information" has become somewhat of an oxymoron in the modern digital age, in that very little information is personal anymore. Data such as medical, financial, and employment records can be accessed by technologically adept people even when such files are not open to public inspection and are protected by laws and regulations. It is not only "hackers" that are causing concerns about the obliteration of privacy, but the expanding surveillance abilities of government and law enforcement agencies is cause for apprehension as well. Surveillance and biometric technologies have advanced exponentially in recent years, making it difficult to avoid seeing images of an Orwellian society looming closer than ever before. If masking your telephone number is tough, hiding your online activity and digital records is nearly impossible.

In addition to the lack of privacy that accompanies modern technology; technological solutions are becoming particularly useful in making it more and more difficult for a person to conceal his or her identity. The enduring conundrum over who can be trusted in cyberspace or in any digital transactions for that matter is being exacerbated by technologies that unearth concealed identities. Weak forms of digital identities are already broadly used in the form of bank account and social security numbers. They provide only limited protection, for it is a simple matter to match them with the individual they represent. As such, concerns about identity and anonymity in the modern digital era are becoming increasingly analyzed, discussed and speculated upon.

### Biometric Identification

*How extraordinary! The richest, longest lived, best protected, most resourceful civilization, with the highest degree of insight into its own technology, is on its way to becoming the most frightened. —A. Wildavsky*

Biometrics is becoming an increasingly popular method of identifying unique human characteristics as a means of authenticating an individual's identity. Where it used to be employed exclusively in crime solving endeavors and high powered corporate or government security, the science of biometrics is quickly becoming about as commonplace as the personal computer.

The science of biometrics is ultimately based on the analysis of distinctive physical traits, such as fingerprints and retinal scans; as well as personal characteristics such as physical, biological and behavioral patterns. Examples of personal characteristics include voice pattern recognition and handwriting analysis. The overall goal of biometrics is to use modern technology to identify individuals, and authenticate their identity, in a more effective and efficient manner. Biometric recognition can be used in the identification mode or the verification mode. In the identification mode, the system identifies a person from the entire population by searching a database for a match. In the verification mode, the biometric system authenticates a person's claimed identity from his or her previously enrolled pattern.

In the business world, biometrics is primarily used as a security measure to prevent unauthorized personnel from accessing confidential data. The use of biometrics in business is based on not what the user knows, or what they carry, but who the user is, some unique characteristic. One method that is becoming more and more popular is keystroke analysis, which authenticates the user based upon their typing characteristics.

Keystroke analysis studies keystroke latency, or the time between successive keystrokes, as well as hold-time characteristic, or the time to press and release a key. These factors are unique to individuals, especially as research has lead from pattern recognition approach such as linear and non-linear distance techniques, z-tests and Bayesian classifiers, to algorithms such as the feed forward multi-layered perceptions algorithm, the radial basis function algorithm, and the generalized regression neural network among others. Using neural network classifiers to perform classification found there is an error rate of only about 12%, suggesting that this approach provides cell phone users with more and better security. Overall, this shows that ability for classification algorithms to correctly discriminate between the majority of users with a relatively good degree of accuracy based on the hold-time of a key.

The data collection, classification and authentication engines would work without inconveniencing the user. The system is best used by users who use cell phones regularly and is not for users with large variations in their handset interactions. In the future, cell phones with built in videoconferencing cameras could adapt facial recognition to strengthen mobile security.

Biometrics is not just used in the business world, but it is becoming a part of home security as well. For example, biometric security systems, like the fingerprint scanner available on the newest computer, is becoming more common for home use. Furthermore, biometric door locks are becoming increasingly popular in home use because they are convenient (no need to fumble around looking for keys) and they are more reliable security-wise than traditional locks. They can be used by scanning one's fingerprints, retina or other parts of the body that make an individual entirely unique using Biometric Access Control.

Even public schools are not immune to the growing biometrics trend, as the scanning of the literal 'student body' is becoming commonplace. Some schools use portable scanners to collect digital images of the students' fingerprints, which need to be updated regularly as the students grow and their fingers change. Biometrics is used for everything from the authentication of new transfer students, to providing the ability to buy lunch in the cafeteria without cash, to checking out books from the library to recording student attendance.

As this trend continues to grow, there is some concern about privacy issues, which are especially sensitive when minors are involved. As reported in the article, *Learning to Live with Biometrics*, Chris Hoofnagle, Associate Director of the Electronic Privacy Information Center in Washington, D.C. believes that fingerprint scanning schools "sets a dark precedent, conditioning students at a young age to embrace the idea of Big Brother-style biometric tracking...If ever there was a generation that would not oppose a government system for universal ID, it's this one".

It is certainly understandable that biometrics would conjure up images of futuristic Orwellian disaster scenarios. However, the main complaint of using fingerprint scanning schools is not about privacy but about a lack of efficiency and convenience. Bob Engen, President of Educational Biometric Technology, suggests that "speed, not security or privacy, seems to be students' biggest concern with the system. The fingerprint-recognition systems tend to run slowly - slower than manually punching in a number, for instance - if a school is using a computer that is more than a few years old. Additionally, large student populations can slow the system since it has to run through every stored image before identifying the best match".

Clearly there are still adjustments to be made for biometrics technology is used pervasively in businesses, homes and schools. However, modern biometrics are machine readable and recordable. Biometric information recorded by machine, and the data linked to biometric observations, can be copied easily, shared quickly and widely, combined, and stored for long periods of time without degrading. Biometrically authenticating identification by machine is highly accurate and, more important, it is highly usable personal information. It allows institutions to collect data and index it to precise and highly accurate human identifiers, making it useful in countless ways.

Currently, businesses are the most prevalent users of biometric technology as compared to homes and schools. However, the popularity of this technology in all realms is continuing to escalate. Does this mean we are headed for a "Big Brother" type of society? Or are we merely looking forward to a more secure, more convenient and more efficient society? Only time will tell, however if the present is any indication of the future, modern technology will soon advance to the point where virtually no human function or interaction will be technology-free with obvious implications for privacy.

### Identity Concealment

*Sacrificing anonymity may be the next generation's price for keeping precious liberty, as prior generations paid in blood.* —Hal Norby

Technologies designed to conceal individual identities are advancing, yet they are still not the point that they provide full protection. The advent of smart cards that generate changing pseudo-identities will facilitate genuine transactional anonymity. 'Blinding' or blind and digital signatures will significantly enhance the protection of privacy. A digital signature is a unique 'key' that provides, if anything, stronger authentication than my written signature. A public key system involves two keys, one public, and the other private. The advantage of a public key system is that if you are able to decrypt the message, you know that it could only have been created by the sender. Modern technology offers a variety of ways of uncoupling verification from unique identity. Validity, authenticity, and eligibility can be determined without having to know a person's name or location. Public policy debates will increasingly focus on when verification with anonymity is, or is not appropriate and on various intermediary mechanisms that offer pseudonymous buffers but not full identification.

Those who are concerned that privacy protection must be strengthened call for government action to require organizations that have confidential information to encode all data and impose electronic barriers against intrusion. This could require government supervision and registry of coding keys. Those who believe that government regulation is not needed argue that voluntary measures by such organizations will be sufficient and point out that, to a great degree, privacy is already compromised since personal information is shared by insurance companies and credit card companies whenever a person uses medical insurance or applies for credit. In these cases, people waive their rights to privacy in order to obtain benefits. Yet there is concern that imposing government interference could lead to the danger of a police state.

Threats of terrorism have, of course, exacerbated the quest for technology that makes identity concealment impossible. This is logical considering that the 9-11 terrorists succeeded in large part because they were able to slip through the system. While the concept of National ID cards emerged long before September 11, 2001, the 9-11 tragedy incited a strong resurgence of demand and controversy.

Recently, the Interpol's Secretary General revealed that cybercriminals have opened two fake Facebook accounts using his name and used them to gather sensitive information. The impersonators were using the Secretary General profile to obtain information on fugitives targeted during an Operation. This again, points out that our world is increasingly connected and networked and therefore also increasingly vulnerable to disruptions caused by intrusions and cyber-attacks. The Secretary General stated,

"Cybercrime is emerging as a very concrete threat. Considering the anonymity of cyberspace, it may in fact be one of the most dangerous criminal threats ever". Interpol benefits from the information gathered and shared by 188 member countries and they are responsible of keeping it safe and of organizing a secure communication network. Among the problems Interpol is facing is the one concerning identity verification and Interpol is currently working on an e-Identification Card - a tool that will be used by staff and law enforcement officials worldwide to prove their identity when accessing Interpol's facilities and its networks and when crossing international borders.

The American Civil Liberties Union cautions that if a National ID card and database were introduced, "the linkage of government databases with corporate databases increases the likelihood that intimate personal information, credit histories, spending habits, unlisted telephone numbers, voting, medical and employment histories-could be easily accessed without a person's knowledge." The American Civil Liberties Union characterizes ID cards as a blatant invasion of privacy and suggests that a computer registry represents a direct violation of Americans' basic civil liberties.

The main concern regarding the introduction of National ID cards centers on the ongoing debate between the rights of privacy and the security of the people. However, apprehensions also exist regarding elitism. According to the Anarchist Federation, "ID is a class issue – the rich will ensure their anonymity by their limited need for the welfare state. Most recently we hear that children of 'celebrities' (which will undoubtedly include well-known politicians!) will be exempted from the Children Index - yet another clear message that ID will not affect everyone in the same way. As well as money, power and influence will give the upper classes anonymity from the state and the private companies who will run identity databases".

## Conclusion

Few things are of more importance to Americans than their rights. The right to privacy has always been held sacred by Americans, but at the same time, the right to have access to knowledge is valued as well. As such, there is a perpetual conundrum relating to identity protection and anonymity in the sense that it is impossible to have it both ways. Protection of privacy and the free distribution of information are innately contradictory ideals. People want the right to know what the government is doing at all times, but the idea of the government knowing what they are doing at all times seems appalling. People are assured that they conduct transactions safely and securely in cyberspace, and yet the amount of information that Internet companies are able to access about their users is phenomenal. Essentially we are living in an age of constant contradictions between what we want technology to do *for* us, and what we fear that technology will do *to* us.

## Privacy Concerns Regarding Electronic Health Records

The Security Journal, Spring 2010

*The way in which information is stored is of no importance.*
*All that matters is the information itself*--Arthur C. Clarke

### Introduction

The increasing use of electronic health records (EHRs) over the past decade or so has brought up numerous concerns regarding the confidentiality of patient data. EHR systems, which allow patient records to be accessed by authorized medical personnel anywhere and at any time, are designed to provide a variety of benefits. The benefits claimed for EHRs are that by being able to quickly and accurately access a person's entire health history, deaths due to medical errors (estimated to be 100,000 a year in the U.S. alone) will be drastically cut, billions of dollars in medical costs will be saved annually, and patient care will be significantly improved.

Clearly, these are very important advantages. However, as with just about everything in life, you have to take the good with the bad; and there are just as many disadvantages to EHRs as there are benefits. The disadvantages stem from not only the lack of privacy and confidentially associated with these records, but also from the inability to properly control who has access to them. While the best intentions would have only qualified medical personnel looking these records, the reality is that interested parties, such as insurance agencies, pharmaceutical marketers, news reporters or anyone else with less than honorable intentions could conceivably hack their way into these files.

### Security Breaches

These concerns are not just based on speculation. Actual security breaches of EHRs have already occurred on numerous occasions. In December of 2006, someone smashed the window of a car belonging to an employee of Providence Health System in Oregon and stole computer backup tapes and disks containing records of 365,000 home health patients.

This incident was just one out of many security breaches of patient records that have occurred in recent years. Others include:

• Wilcox Memorial Hospital in Lihue, HI, lost a thumb-sized data drive with information on 130,000 former and current patients.

• Backup tapes containing information on 57,000 enrollees of Blue Cross Blue Shield of Arizona were stolen in a burglary of a managed care company that worked for the insurer.

• A hacker broke into a server and nabbed 42,000 patient records at the health center of Colorado University in Boulder.

• Kaiser Foundation Health Plan was fined $200,000 by the state of California for posting information on approximately 150 patients— without their permission
    —on a public website.

Clearly security and privacy issues are a major concern in regard to electronic health records. But at the same time, for many people the idea of saving tremendous amounts of lives and money makes EHRs seem worth the risk. For mental health professionals the dilemma is especially tricky because many mental health patients would be devastated if their personal information somehow got into the wrong hands and was made public. For example, in mental health, effective treatment is built on a foundation of trust between clients and providers and often requires the sharing of sensitive information. Insurance and employment discrimination, stigma, and financial repercussions can result when data are shared inappropriately.

Creating security systems that will ensure privacy and confidentiality is of the utmost concern for mental health professionals and their patients. Most people do not want outsiders, like drug companies, to be able to gain access to their health records, with the exception of the need to perform medical research. People are aware that computer security breaches happen all the time and they do not want to fall victim to this problem. This is quite understandable considering that the annual loss due to computer crime was estimated to be $67.2 billion for U.S. organizations, according to a 2005 Federal Bureau of Investigation (FBI) survey. Further the annual Computer Security Institute and FBI Computer Crime and Security Survey for 2005 for example, reported that only 20 percent of business who suffered serious computer security breaches reported the incident to law enforcement. The key reason cited for not reporting intrusions to law enforcement, according to the Report, is the concern for negative publicity.

### Paper Records

Or course, there is the argument as well that non-electronic records (traditional paper records) could be just as vulnerable, if not more vulnerable to invasion of privacy. After all, at least in electronic storage systems there are security measure taken to keep them safe. Paper records usually have no more protection than a lock on a filing cabinet. On the other hand, security leaks involving paper records are much smaller in scale than electronic leaks. An individual who gets his or her hands on a stack of paper files is not going to gain anywhere near the amount of data that someone who is able to hack a computer system is going to get.

### Privacy measures for EHRs

Measures are, of course, being taken to ensure the privacy of electronic medical records. According to a recent article in *Healthcare Technology Management* (2009) former U.S. Health and Human Services (HHS) Secretary Mike Leavitt outlined eight "privacy principles" in a speech given at the Nationwide Health Information Network Forum in December, 2008. These included the following:

1.      Consumers should be provided with a simple and timely means to access and obtain their personal health information in a readable format.

2.      Consumers should be provided with a timely means to dispute the accuracy or integrity of their personal identifiable health information, and to have erroneous information corrected or to have a dispute documented if their requests are denied.

3.      Consumers should be able to add to and amend personal health information in products controlled by them such as personal health records (PHR).

4.      Consumers should have information about the policies and practices related to the collection, use and disclosure of their personal information.

According to *Healthcare Technology Management* (2009), the other principles focused on issues of access permission, data integrity, and accountability. Leavitt also provided information about how to avoid security breaches. All of this information is critically important considering that President Obama has announced that he plans to have all health records electronically stored within the next five years.

The U.S. Department of Health and Human Services has been working hard to make sure that the healthcare profession is armed and ready to meet this challenge with as few threats to privacy as possible. In a report titled "Enhancing Protections for Uses of Health Data: A Stewardship Framework" (2007), and HHS advisory committee known as the National Committee on Vital and Health Statistics (NCVHS) made a formal request for certain frameworks and terminologies in the Health Insurance Portability and Accountability Act (HIPAA) to be updated to suit current needs. Of particular concern regarding terminology was the use of the phrase "secondary uses." The phrase is currently used to refer to medical researchers, but the vagueness of the phrase, and the loopholes it can open, have caused the NCVHS to submit to the Secretary of Health and Human Services (HHS) that it "urges abandoning this ill-defined and overly broad term in favor of naming each use of health data—for example, reporting communicable diseases to public health, or informing the quality improvement process".

An additional concern that has reformers fired up is the fact that there are so many incompatible systems. The current plethora of incompatible systems—personal and physician health records, hospital records, lab tests, imaging, payer data, web portals—makes it difficult to guarantee information security. Health reform should empower patients to easily access and control release of their own health records, but it is hard for patients to assemble and share complete records from multiple sources.

## Policy Changes

Clearly, regulations are changing due to rapid changes in technology that have allowed EHRs to be a viable possibility. It is likely that legislation and other reform efforts will continue as more and more electronic records become accessible. However, while Congress should establish a strong framework for health privacy and security, it must avoid a "one size fits all" approach that treats all actors that hold personal health information the same. The complexity and diversity of entities connected through health information exchange, and their very different roles and different relationships to consumers, require precisely tailored policy solutions that are context and role-based and flexible enough to both encourage and respond to innovation.

As it is now, legislation governing EHRs already varies significantly between jurisdictions. Various jurisdictions have established legal requirements of varying degrees of clarity on what information is to be recorded, how long information is to be retained, and what its content should be contained. Given that paper records have been the predominant source of patient information for centuries, it is understandable that new legislation needs to be enacted to account for this major overhaul of traditional systems. As such, the rules, formulated for paper records, are undergoing modification and extension to adjust to the new opportunities, cost considerations, and risks introduced by the digitization of the medical record. Nonetheless, online practice does not diminish patient protections; at least until the law becomes clearer, all applicable laws should be applied to electronic records.

As of 2004, only thirteen percent of patient medical records were electronically accessible, primarily due to the cost of transference. However with President Obama encouraging 100 percent transference within the next five years, drastic changes in legislation, practice and security are bound to occur.

### Privacy and Security Measures

The number of security breaches among medical institutions both large and small, continues to increase despite the plethora of measures continuously devised to combat intrusion. Even the U.S. government, whom one would think would have the most air tight computer systems in the world, has been vulnerable to hackers on numerous occasions. According to the Privacy Rights Clearinghouse, there have been over 500 security breaches since 2005, many involving the most respected organizations in the United States. This is evidence that security breaches can affect anyone at any time, no matter how powerful and organization may seem.

Many people tend to think of the corporate healthcare giants of being somewhat immune to routine security problems, simply because they have the financial resources at hand to combat them. However, it is becoming increasingly clear that 'money can't buy security' if the impenetrable device does not yet exist or being developed. It seems with every new innovation in security comes a new innovation in security breaching, and the race on both sides of the fence seems infinite. Security measures and forensic techniques are becoming more advanced, but have yet to surpass the rate at which intrusion technology is growing.

Currently, there are numerous perspectives as to what privacy protection and security measures should entail. There is a fairly comprehensive list of security measures that few could argue with and the following need to include:

1.    Accountability-proof that an intended transaction indeed took place.

2.    Confidentiality-protection of confidential information from an eavesdropper.

3.    Integrity-assurance that the information sent is the same as the information received.

4.    Authority-assurance that those who request data or information are authorized to do so.

5.    Authenticity-assurance that each party is who they say they are.

Effective security program must support the needs of the business without imposing excessive cost or inconvenience to users. Overly stringent security controls can lead to productivity loss among workers who wrestle with time-consuming processes. Overly restrictive controls undermine efficacy, as in the example of a demanding password policy that leads users to write passwords on sticky notes pasted to their monitors. Unnecessarily stringent controls increase costs of technical support while delivering little incremental risk reduction.

By combining several non-intrusive security measures that work together, an organization can achieve dramatic risk reduction without more costly and infringing controls. A less demanding password requirement, combined with the right network configuration, on top of a properly communicated and enforced human resources policy, can together provide greater security than a single more powerful control.

The increased demand for online data has, in turn, created a variety of new security problems for statistical enterprises, in particular, what are known as "tracker attacks". Anomaly detection techniques capture both known intrusions and unknown intrusions if the intrusions demonstrate a significant deviation from a norm profile. Existing anomaly detection techniques differ mainly in the representation of a norm profile and the inference of intrusions using the norm profile. The disclosure limitation has been the primary focus of concern; however, more attention needs to be paid to software devices that prevent illegal access to critical, confidential information. Ultimately, although secure software methodologies will continue to improve, bugs and vulnerabilities in computer and network systems are inevitable due to the difficulty in managing the complexity of such large-scale systems during their specification, design, implementation, and installation.

## Conclusion

As progressively more patient health records continue to become electronic instead of paper-based, the need for advanced security systems to protect privacy will grow exponentially. People's fears about their personal, private information becoming available to outside parties are, unfortunately, not unfounded. Security breaches continue to occur despite elaborate measures being taken to prevent them. In the end, the issue comes down to what is more important – saving hundreds of thousands of lives and dollars through an integrated online system, or taking the chance of private medical records becoming public fodder. It may very well be that people will have to be willing to sacrifice some of their feelings of personal security in order to serve the greater good.

## Security in Digital Health

The Security Journal, Summer 2010

### Introduction

More healthcare data is becoming digital every day. Over the next several years, the amount of digital personal medical information online will grow exponentially. The increase of new opportunities for hackers to expose personal data, unlike financial data, this personal data could cause permanent damage to privacy of individuals. The U.S. Department of Health and Human Services (HHS) has set a deadline of 2015 for healthcare facilities to being using electronic health records (EHRs), thereby ushering in the digitalization of all patient information. As patient data is aggregated on health networks, it becomes a bigger target for those who want to steal it and exploit it on the Internet.

However, a number of barriers must fall before electronic patient records, personnel data and other sensitive data becomes the norm. This is because when it comes to digital healthcare, even just one security breach can be extremely costly, both from damaging the medical facility's reputation as well as actual damage to the bottom line. The expenses of security breaches must therefore be balanced out with the costs of in-house security experts to manage the implementation and upkeep of data security systems. Although an in-house security staff can be expensive, and outsourcing may seem on the surface to be the most reasonable option for obtaining cost-effective security, the safest way to keep digital health data secure is to manage it in-house.

From the perspective of the patient's, there are widespread concerns about the use and abuse of digital healthcare data. Standards and procedures have not kept up with the speedy availability of information. Digitalized medical databases cross state lines, though laws regulating the confidentiality of patient's medical records vary widely. In particular, many Americans are afraid that their personal and "private" medical information might be used against them to limit their insurance coverage or ruin their chances of a getting a job or promotion. Overall, there is a widespread concern about what digital health records mean for privacy in general. Unfortunately, these concerns are not unfounded. For example, very recently, "At Marysville's Rideout Memorial Hospital, 17 security guards rifled through the personal health data of 33 patients, using computers to peer into what should have been private and protected electronic health records".

Sadly, this is not an uncommon type of incident. Between 2000 and 2007, security breaches in the healthcare industry constituted 11 percent of all nationwide electronic security breaches, and almost half of those occurred in hospitals. More recently, it has been reported that "Between September 22, 2009, and February 15, 2010, at least 47 instances of breaches of unsecured protected health information occurred in the United States, each affecting at least 500 individuals with one affecting more than 500,000" (Goedert, 2010).

**Trends**

A major part of the problem stems from a lack of compliance to regulation. A 2010 Identity Force survey, involving over 200 compliance experts across the United States, revealed that despite increases in laws and regulations relating to compliance with security standards, the number of security violation incidents in the healthcare industry continues to rise. The CEO of Identity Force, Steven Bearak had this to say:

"It turns out that addressing the problems of data breaches and medical identity theft is proving more complex and time-consuming than hospitals counted on. It's interesting that many compliance experts now call into question whether or not even the new Healthcare Reform Law will ultimately help on these issues. We are simply copying, digitizing and disseminating personal information faster than we can control it."

There is a "prima facie" obligation on part of the healthcare institution to arrange its information structures in such a way as to permit the establishment of data-need profiles, based on specialty and information needs, which would allow the system to flag (or exclude) unacceptable access to parts of the patient record. In this way, the institution would guarantee that the genuine data needs of the relevant health care professionals are met irrespective of whatever changes might occur in the professional environment, while at the same time fulfilling its obligation of privacy and confidentiality towards its patients. However, at the same time, the institution, while fulfilling its duties towards its patients, must be careful not to erect unreasonable access barriers in this regard since its ability, its functioning, and the ability of the professionals to access it is all rooted in a need for faster, more efficient flow of information. Ultimately, the challenge is to find the right balance between too much restriction and not enough.

### Importance of Protecting Data

The importance of security measures to protect data is extremely important in the healthcare industry because of the sensitive nature of patient information. Even the largest healthcare organizations do not always have the most effective data security systems. Security breaches can affect anyone at any time, no matter how powerful and organization may seem to be. Yet at the same time, it is important to recognize that being too cautious can defeat the entire purpose of digital records, which are designed to provide instant answers to critical problems without delay.

The basic premise that underlies an examination of the information-relative rights and obligations of a health care institution, whether from an ethical or a legal perspective, is that health care institutions are entities to whom ethical and legal considerations apply in the first place. Patient records in general and electronic patient records in particular, occupy a special place in health care decision-making. Operationally, this results from the fact that the data which constitute the records are the basis of the information that is used in health care decision-making about patients. It is in fact, the inherent value of medical records that makes them so vulnerable to misuse.

### Patient Concerns

Patient records are not, however, the only concern when it comes to privacy and ethical issues related to e-health. According to Sanchez-Abril and Cava (2008), online health networking websites create problems of their own:

> "The many privacy risks posed by online health networking involve a complex web of real-life and cyber-relationships, questionable duties to the cyber-patient, and the technological capabilities to widely and permanently publish another's private information. Consequently, health networking privacy breaches can have several perpetrators: a malevolent blabbermouth, a mercenary web operator, a medical identity thief, or even an impulsive cyber-patient with a false sense of security" (p. 246).

Few things are of more concern to patients than the privacy of their medical information. People generally do not want other people talking about their medical problems, but they want to discuss them with others online. Unfortunately, despite anonymity, privacy is difficult to maintain. To avoid the types of problems that can occur in networking situations it is recommended that Internet Service Providers offer cyber-patients the following tools:

- Providing clear language defining anonymity and educating users about the level of protection against identification on the website and how it is affected by altering privacy settings;

- Providing users with available technological tools to mask their identities in appropriate and clearly-defined circumstances (anonymization through technology);

- Assuring third parties will not gain access to identified information; and

- Indicating the precise circumstances (such as a government-issued subpoena) under which the ISP will provide a party with user information, whether content-related or otherwise.

It is undeniable that the Internet has created a number of opportunities for society. As a tool of commerce, the Internet offers consumers easy access to a wide range of goods and services and offers businesses a new medium through which to market and sell their products and services. As a tool of communication, the Internet also provides individuals with access to a broad array of information. While providing these and other benefits, the widespread use of the Internet has also raised some cause for concern, especially as far as the protection of personal information is concerned.

### Identity Theft

Ultimately, effective data security is about more than just finding and implementing the best technology. The true challenge lies in ensuring compliance with the policies that regulate access to medical information. The greatest data security systems in the world will not be any more effective than the most well-intentioned laws if people do not respect them. People with mal-intent and the know-how to take advantage of technology will be inclined to do so. As such, identity theft is another major concern regarding digital medical records. In fact, a recent report by Javelin Strategy & Research revealed that "data theft related to exposure of medical records rose in one year more than 100%, from 3% in 2008 to 7%, or 275,000 cases, last year". Once the cyber thieves obtain this information, they can use it for anything from standard identity theft purchases to major insurance scams.

While medical identity theft is still in its early stages, a recent article in U.S. News and World Reports indicates that it is an ever growing problem:

> "Medical identity theft currently accounts for just 3 percent of identity theft crimes, or 249,000 of the estimated 8.3 million people who had their identities lifted in 2005, according to the Federal Trade Commission. But as the push toward electronic medical records gains momentum, privacy experts worry those numbers may grow substantially. They're concerned that as doctors and hospitals switch from paper records to EMRs, as they're called, it may become easier for people to gain unauthorized access to sensitive patient information on a large scale. In addition, Microsoft, Revolution Health, and, just this week, Google have announced they're developing services that will allow consumers to store their health information online. Consumers may not even know their records have been compromised".

### Digitized Health Information

President Barack Obama recently instituted the "HITECH Act" to help solve the types of problems associated with digital medical records and online health information. The Federal Government has made a commitment to Health Information Technologies (HIT) by creating the Office of the National Coordinator for HIT. Limited funding has been devoted to establishing Federal HIT standards and supporting the development of a national health information network. A recent Executive order directs Federal agencies to promote HIT as a means for improving the quality, efficiency, and transparency in the health care system. In addition, bipartisan legislation aimed at furthering the use of HIT and the exchange of digitized health information has recently been passed in Congress.

Despite the public policy attention being paid to the use of technology in health care settings, implementing a national HIT and Health Information Exchange (HIE) infrastructure will be a complex and difficult task. The significant number and diversity of stakeholders, the legal, privacy, and security issues, the lack of agreed on standards for data exchange and storage, the source of funding for these technologies, the uncertain return on investment, and the uneven distribution of the return present persistent issues that must be addressed if these technologies are going to achieve wide-scale improvements in our health care system.

### Conclusion

On one level, electronic patient records are merely electronically based devices that fulfill the same role as their previous material counterparts, much as is the case with computer-based medical imaging, telemetry, etc. In that sense, the development of electronic patient records parallels developments in the latter areas. They are merely more easily accessible than the old-style material records, more easily manipulated, and potentially give a much more complete representation of their subjects than could previously be achieved. However, as in so many instances, what on the surface appears to be merely a change in technology hides a much more profound development in the nature and status of what was thus changed.

In this particular instance, the technological switch from paper-based records to electronically based records facilitated a functional development that radically altered the relationship between the subject of the record and the record itself, and widely expanded the uses to which the record could be put. It is this change in relationship, and this expansion of usage that gave rise to a whole series of ethical problems: problems that center on issues that encompass everything from invasion of privacy to identity theft to the professional misuse of confidential information. There is no blanket solution to address these complex issues, and in all likelihood, there never will be. However, on the whole, the benefits of modern technology in the healthcare profession seem to outweigh the negatives.

## Why Every Organization Should Have a Data Protection Strategy

The Security Journal, Winter 2010

### Introduction

The importance of security measures to protect data is often underestimated, especially in smaller companies. As such, many small companies tend to ignore or avoid data protection, assuming that it is only something with which multinational conglomerates must be concerned. Even medium sized and large companies often take data security for granted, implementing only the most basic of security measures. Furthermore, even the largest corporations do not always have the most effective data security systems. For example, according to the Privacy Rights Clearinghouse, there have been over 500 security breaches since 2005, many involving the most respected organizations in the United States. This is evidence that security breaches can affect anyone at any time, no matter how powerful an organization may seem to be.

Many people tend to think of the corporate giants of being somewhat immune to routine security problems, simply because they have the financial resources at hand to combat them. However, it is becoming increasingly clear that "money can't buy security" if the impenetrable device does not yet exist. There is an old axiom that says "whatever man can conceive, he can deceive." Many times, security boils down to; what do you gain and what do you lose by implementing a given solution. It seems with every new innovation in security comes a new innovation in security breaching, and the race on both sides of the fence seems infinite. Therefore, the most important step that organizations of all types and sizes can take is to create a strategic security plan that takes into account all possible scenarios.

In many organizations, because so many computer users and system administrators are not sufficiently aware of the vulnerabilities of their systems, they are not demanding more secure systems. Moreover, organizational leaders do not typically view security as a top priority and find few incentives to purchase products than can make IT operations significantly more secure. It is important to have top management buy in and strategically align IT security goals with business goals. IT security needs to be thought of in terms of business impact. Even though software and hardware developers are continuously creating new and better security technologies, it often takes a very long time for those advances to find their way into organizational systems.

In reality, every organization, no matter how large or small, should have a strategic and tactical plan in place for protecting the organization's data and personal information. Organizations must identify the risks to the most critical assets. The National Strategy to Secure Cyberspace is the federal data protection strategy for the United States. Ultimately, this strategy should serve as a model for all organizations.

### Strategy Development

Strategy determines the basic long-term goals and objectives of an enterprise, the adoption of courses of action, and the allocation of resources necessary for carrying out these goals. Strategy defines the alignment of facilitation among organizations' missions, visions, business strategies, and strategic initiatives. Through effective strategy development, organizational leaders can ensure a logical course of action and promote coherent decision making across all levels and functions of an enterprise. Organizational leaders must think in terms of both means and ends when considering goals and decisions of a data protection strategy.

One of the main principles of developing an information security strategy is to be proactive. The concern that network operation centers (NOCs) and security operation centers (SOCs) are more focused on business impact as opposed to hardware and software impacts. The solution to this problem lies in taking proactive security measures and developing an effective program for handling security incidents. Proactive security measures include; developing policy, performing a thorough analysis of security vulnerabilities and implementing measures to secure those vulnerabilities.

In the terms of operations security, there exists in most organizations, fragmentation within the company, a lack of effective communication, and an unresponsive chain of command. In this regard, only a few organizations have formal programs to address personnel or operations security. The pertinent question that needs to be asked is: What would constitute an ideal model for security governance? The answer to this question is a three-point solution plan that consists of (1) the development of an integrated security function; (2) appointing a Senior Manager of Security to oversee the integration process; and (3) creating an increased awareness of the importance of security within the firm.

The National Strategy to Secure Cyberspace has developed a proactive strategy that organizations should strive to emulate. The National Strategy focuses on the following five priorities:

4.     Priority I: A National Cyberspace Security Response System

5.     Priority II: A National Cyberspace Security Threat and Vulnerability Reduction Program

6.     Priority III: A National Cyberspace Security Awareness and Training Program

7.     Priority IV: Securing Governments' Cyberspace

8.      Priority V: National Security and International Cyberspace Security Cooperation  (National Strategy to Secure Cyberspace, 2003)

Although these areas of focus are designed to address a larger scope than that of the average organization, there are elements of each of these sections of the National Strategy that could be readily applied to the private sector. For example, one of the objectives of Priority I is to, "Provide for the development of tactical and strategic analysis of cyber attacks and vulnerability assessments". This should be an action that is part of every organization's data protection strategy. The same holds true for this objective from Priority II:

"Understand infrastructure inter-dependencies and improve the physical security of cyber systems and telecommunications". Priority III and Priority IV also include goals that should be applied to organizational data protection strategies, such as "Foster adequate training and education programs to support the [organization's] cybersecurity needs"; and "Authenticate and maintain authorized users of...cyber systems". Priority V contains relevant applications to the private sector as well, including "Improve capabilities for attack attribution and response"; and "Foster the establishment of...'watch-and-warning' networks to detect and prevent cyber attacks as they emerge".

Organizations of all shapes and sizes need to implement formal data protection strategies that include these and other objectives as outlined by the National Strategy to Secure Cyberspace. New threats and vulnerabilities are constantly emerging and so too must organizations change to deal with these innovative security threats.

## Dealing with Change

The United States (and the rest of the world) is undergoing dramatic change, due in great part to the dramatic transformations brought about by new information technologies. The technical changes include advances in how information is collected, stored, processed, and communicated. While the speed with which these processes have taken place has increased significantly, the costs for propagating and storing information have decreased dramatically as well. The implementation of these capabilities has vastly increased our communications and related functions. These changes have been rapid, and more are on the way. Advanced information technologies will continue to fundamentally alter how people interact, and in ways that cannot be predicted.

Unfortunately, along with these widespread changes have come vast opportunities for crime and corruption. The growing numbers of wireless applications and increased network access are making it easier than ever for unauthorized users to view confidential information. The criminal activity has increased simultaneously with technological advances, and for businesses, security leaks have become a common problem.

Security threats include everything from identity theft to corporate embezzlement to hacking to invasion of privacy. With security and privacy controls becoming more internalized within most organizations, a need naturally arises for more advanced internal management of security risks. This means not only increasing and enhancing training programs that are ongoing and up to date, but also providing written documentation in the form of a brochure or computer generated handout.

Due to constantly changing information technology it would not be cost effective to merely have a brochure explaining the security strategy printed up and distributed to employees. The formal document must be something that can be easily altered and then printed and distributed easily, inexpensively and efficiently.

Therefore an in-house generated document that can be updated at will is the best possible supplement to ongoing training.

Clearly, effective security involves more than obtaining the right technology. The real challenge is assuring its effectiveness, surrounding it with policies and practices that reduce risk, and addressing a changing environment. Traditional approaches to information security offer little help considering that the basic principles of encryption, authentication, and security that provide the foundation of most Internet security systems are available in a variety of ever evolving forms. Good security must be a dynamic process that addresses a constantly changing environment. This requires a steady flow of information and analysis around emerging security issues, to protect against new threats before it is too late. Even risk transfer mechanisms must be examined regularly to ensure that new threats are covered. Policies, practices, configurations must be updated dynamically to remain relevant. This is a complex, long-term process that requires specific attention to not only systems management, but employee management as well. Security strategies are of little value if they do not involve people, technology is not always the panacea. One must continually educate one's workforce.

## Organizational Management of Security Strategy Implementation

As with any new organizational strategy, employees must be completely on board with maximizing the new system. Employee attitudes and behaviors toward data security can be based on the perceived social norms of the organization, and what they believe to be shared standards for acceptable and unacceptable behavior in the workplace. By creating an information protection strategy that will respond effectively to changing conditions, will help map out the importance of protecting data. This will assist management in gaining support from personnel.

Involving employees in the strategy development and implementation process is also a surefire way to get employees to support the change. When employees are directly involved in the decision-making and strategic planning processes, they are more likely to take ownership of the plan, and in turn, are more likely to support it because they have a personal investment in seeing it succeed. These strategies and policies are all part of risk control and should all be tied back to business objectives. Strategic assets, along with the development and exploitation of strategic advantages, are often rooted strongly in the people of an organization. Thus, the effective management of human resources must come to play an integral role, in understanding the major issues involved in the strategic implementation process of the data security plan.

In addition to providing a general overview of legal and ethical problems associated with information technology, this adds a slight 'twist' by approaching the subject from a 'corporate culture' standpoint. A great deal of time must be spent reinforcing the point that traditional ideas about security and other IT issues are no longer optional, they are imperative. For example, to employees the need for strong passwords and other measures to control access no longer seems like an annoyance, but an expectation. Now when the security team explains to employees why it is against company policy to share passwords or post them near workstations, they listen and might even ask where the protections are if they are absent.

There have undeniably been some dramatic changes in IT over the last four decades. Changes that were considered only science fiction and simply not possible a few years ago are reality today. The technology to communicate with anyone anywhere is truly a phenomenon that is changing the world in which we live. Businesses are having to react in real time and have to build for a future that will be in constant change. What has not evolved as quickly are the people issues that surround the development and use of IT in a secure and prudent fashion. Many companies still struggle with poorly defined security problems and even more poorly defined solutions. There are underlying and fundamental issues that many organizations have yet to come to grips with, including the fact that change is still not easy, that people still do not want to change, and that better methods for change management, to get people to understand the importance of data security--are still needed.

## An Organization Strategy

Empower employees with the stewardship of treating all data, both electronic and paper as if it contained your own personal data. You do not have to be an information security or privacy expert to apply your knowledge from life experiences with safeguarding your own sensitive information. From the moment the employees' workday begins, the awareness of the critical importance of protecting information begins. It begins with security screening to enter our workplace and is followed by utilizing multiple computer passwords to access e-files and software.

## Information must be protected

Each time a breach of information security occurs or the loss a laptop with sensitive information is publicized, renewed interest or lingering questions about our readiness to protect information come to light. The easiest way to steal corporate data is to steal a company's laptop; all of your corporate laptops need to be encrypted. For generations, people have defended and protected their own property and privacy using locks, fences, signatures, or seals. People are incredibly smart and can learn some extremely complicated things. We need to discover how to transfer expertise from the people who do understand information protection to those that have been entrusted with the most private and personal information to this agency. Businesses need to have a data classification system, this is one of those items that is easy to talk about, but hard to implement. A data classification system is hard to implement because it involves data owners and requires lots of planning prior to implementation.

## People make the difference

While the information environment has introduced a new set of problems, the issue is not with the new processes or technology but with the human use and misuse of the processes and capabilities. The search for solutions must incorporate an increased awareness of the human behavioral dimension of protecting information and privacy for both the customer and employees of the organization. Information security and privacy solutions must be founded upon such an understanding, since people are both the source of and the solution to the problem. It is impossible to defend against every threat; the best solution is always to have a combination of people, processes and technology.

## Situational awareness

Information Security and Privacy Officers alone cannot secure the organization enterprise—reducing risk requires an unprecedented, active cooperation from each person. Empowering employees in the role of protecting information is the right approach for keeping all information secure. Employees are always one of our weakest links in data security, but just imagine if you could turn it around and move just 10% of workforce into a higher state of security awareness. An organization with a workforce of 250,000 personnel, if you could just get 10% to be more proactive, you would have the leverage of 25,000 additional people/sets of eyes helping you every day.

## Framework

The reliance on information will only continue to grow in the years ahead. A strong effective secure enterprise should combine people, process, and technology to maintain balance and defense-in-depth protection of information. Establishing the following framework will help us achieve a more secure enterprise from the top down.

1. Develop and implement an Information Protection Strategy that provides a framework for protecting an organization infrastructure as an essential part of its mission and way of life. The strategy should include how all employees handle organizational information in the performance appraisal process.

2. Design information systems in a way that maintains confidentiality and include stakeholders in the design, development, and activation state.

3. Educate and train employees to include outreach to the organization's customers regarding rights and responsibilities and disciplinary sanctions for misuse of information to include three keys elements for success: repetitive, make it personal, and make it fun.

## Conclusion

At the present time, readily available technology based network security components such as Firewalls, Anti-Virus programs and Intrusion Detection Systems (IDS's) are unable to effectively combat the vast range of malicious intrusion attacks perpetrated on computer networks and systems. This is where the need for proactive security measures comes into play and becomes a necessary part of making and keeping information technology secure.

There is a great deal more to effective security than choosing the right hardware and software. One of the most important security components available is the employee taking a proactive role in forging a more secure enterprise. People know to protect their wallets and checkbooks, as this provides a quick way for thieves to access money. Instead of protecting wallets and checkbooks, employees can to be proactive in protecting files and electronic records. Effective security is about properly managing both policy and people. Concerns about computer security are far from over, as statistics continue to demonstrate, the threats are increasing every day. Therefore, it is imperative that companies of all shapes and sizes design, implement and distribute a formal strategic data protection plan that is solid, but can also easily adapt to change.

## Telework and Mobile Computing Security Concerns and Risks

The Security Journal, Summer 2010

As the risks and technical challenges of telework and mobile computing multiply, so too does the security and cost to organizations for protecting data in this complex environment. Telework and mobile computing have taken the advantages and enjoyed the many benefits of today's high-capacity broadband networks. Although many of the world's most successful companies have already embraced the virtual workplace, one might think that because of the role telework plays as a motivator, morale booster, and environmentally friendly alternative, everyone who is eligible to telework would be doing so. The risks, the technical challenges, and the cost of replicating the secure office environment are, however, no small feat. The proliferation of mobile devices for today's mobile workforce has exposed a vast amount of confidential company data outside the physical office. The greater reliance on telework and mobile computing raises numerous information security risks and concerns, including everything from breaches of confidentiality, increased opportunities for unauthorized viewing of data, data theft and leakage, and identity theft. Is the risk of cybercrime to the mobile workforce worth the cost of stolen data in the long run?

More than a decade ago, analysts predicted that online communication would create teams or task groups characterized by increased participation, more egalitarian participation, more ideas, and less centralized leadership. Some who proffered opinions also believed that limited social presence may encourage people to communicate more freely and creatively than they might in person.

The anonymity of the Internet was highlighted as enabling people to develop more multiplex relationships. Connections on the Internet found that many on line ties do meet most of the criteria for strong ties. At that time, telework was touted as a future trend, as the growing number of teleworkers included many salespeople, managers, professionals, and support personnel. On the other hand managerial resistance to telework proliferated in the workplace, and it was frequently cited as a major problem, as fears that professionals would suffer because of less visibility in the organization. It was argued that telework curtail informal communication even though this might be mediated by employee status or organizational support. It was also found that managers maintained tight control over teleworkers, suggesting that communication remained roughly the same.

Teleworking involves doing most of one's work at home or at a remote location, but the concept is also being incorporated into a broader idea of the so-called mobile employee. The "mobile employee" is an employee who makes use of a laptop, cell phone, or handheld device at a remote location, often gaining access to the company network or database. When the precepts of telework were first established more than a decade ago, it was also a time in which many of the limitations of telework were established. As Friedman & Hoffman (2008) notes, today close to 81 percent of executives in companies around the world have a mobile device, and nearly 75 percent of the U.S. workforce will soon have mobile capabilities in terms of access to company networks.

While inappropriate for some people, tasks, and organizations, telework appears suitable for jobs that require the interpretation, communication, and manipulation of information, making the teleworker a worker with specialized knowledge. Thus, in many respects, telework can shape those firms that are primarily knowledge-intensive. In such organizations, the capability to exploit existing knowledge and to create new knowledge is what gives a firm a sustainable competitive advantage. Based on the belief that knowledge is the main source of competitive advantage of the knowledge-intensive firms, the potential impact of telework and mobile computing could have deleterious effect on the organization if its knowledge and data are not protected in a telework or mobile computing environment.

### Security Concerns

In each case of security, a company must weigh the benefits of remote access against the inherent dangers. It is now clear the perimeter-based model that supports traditional IT security is out of step in the more complex remote environment. As IT departments are told to let down the drawbridges and allow remote workers to access enterprise data, this has extended the enterprise. Any time an executive takes his laptop home and uses the Internet from home, or a mobile worker messes with the configuration of his or her computer in a hotel room, the connection could contract a virus that moves quickly through the company server, causing enormous damage. The very emergence of the 3G phone means that every phone has an Internet protocol address complete with wide-open ports waiting for someone to take it over. Such open-ended access could add up to millions of conduits into the corporate network, which in turn could equal millions of new ways for the mischievous or downright malicious to gain access to critical company data. Often most companies deal with security matters only when a problem emerges, which is already too late. The reactive mindset of organization security, therefore, needs to be altered, or new technologies that cancel out the negative effects of the mindset need to be in place. However, no matter how private and closed an organization's network is, a risk exists.

New technology is making the line between office and road disappear. New developments in database and software application companies allow people on the road to check on customer orders and corporate inventories before heading to a meeting. Today's high-capacity broadband networks help people work productively from anywhere. As jobs continue to shift from manufacturing toward knowledge-based services, workers, managers, and society at large increasingly realize the benefits of telework with more and more workers opting for its various forms. Mobile corporate computing is coming to the masses primarily because of superfast wireless networks, innovative communications software, and a host of relatively cheap devices from notebook to palm-size handhelds. Many of the top companies are distinguished by their embrace of anytime-anywhere computing. Companies have adopted mobile computing because employees tend to work more for the same pay, and the company can slash overhead at headquarters. One reason for this growth is that mobile computing is becoming more secure and concerns over lost equipment and unauthorized access, a brake on telework development over the past decade, are now disappearing. Disabling devices, firewalls, and other security devices now have allowed even some companies to allow remote access to customer information. Further, thumb drive devices as small as key fobs which can plug into PCs anywhere receive and display regularly updated security codes.

The need for information of mobile users will drive the need of mobility computing. In the case of healthcare, where healthcare is an information-intense field, healthcare providers are now using handheld devices and have found that using such devices help them with their professional confidence. The information derived from handheld devices has resulted in significant gains in self-perceived confidence in clinical decision making. Even now mobile phones provide new ways to gather information both manually and automatically over wide areas.

### Security Risks

While the lines between telework and mobile employment are often blurred, both are part of a trend that brings with it enormous security risks. Telework continues to develop although a number of barriers persist. In the area of federal government employees, a gap exists between those eligible to telework and those actually teleworking. It may be that some of the requirements of telework, such as remote access to an organization's IT infrastructures, still makes managers nervous, and that might well be true in the context of a federal agency. One way to circumvent barriers is to adopt an "opt-in" policy according to which every employee in the organization is made "telework eligible" even if not ready to do so. It is also essential that the technological parameters of telework be carefully framed, as defining specific, authorized-user devices and connections make safe and managed telework easier to accomplish. It is only through such a plan that data security issues can be given full consideration. Top-down managerial support as well as training is also essential for carrying out this plan.

The popular press and increasingly academic literature is replete with reports on malicious attacks on company systems, raising security concerns for all IT systems, especially when remote access is easy. Reports on a widespread injection of malware into the SONY Playstation Web site in July 2006 were typical of attacks by malware taking place at breakneck speed. The malware was injected into the system through a fake security scan offer, which actually downloaded malware if a user pressed "yes." For the untrained remote user, this social-engineering scare tactic has become increasingly common among online criminals. The fact that the virus spread through an unsuspecting fan pressing yes to a security offer emphasized the need for remote user education.

Recent losses and thefts of laptops containing confidential employee and customer data have proved extremely costly and embarrassing to leading companies and government agencies. Hackers see that so-called mobile or enterprise employees are an excellent way of gaining access to corporate networks. They have noticed—even if many companies have not—that mobile devices typically are both the most vulnerable computers in the enterprise and the least defended. One of the main problems is that mobile systems connect to the Internet or shared networks directly, thus bypassing the corporate defenses. While employees in the office work inside impervious LANs, mobile workers use Wi-Fi hotspots that are vulnerable to many types of interception and spoofing. It is also true that loss or theft of company laptops is common and widespread. Many companies spend millions to secure their on-site computers, but then take the view that privately owned laptops and mobile telephone security issues are the individual's concern. There are many types of security threats to mobile devices. For purpose of assessing and setting priorities, the threats can be grouped in the seven categories:

1.  Malware
2.  Phishing and social engineering
3.  Direct attacks by hackers
4.  Data communications interception and spoofing
5.  Loss and theft of devices
6.  Malicious insider actions
7.  User policy violations

Malware itself has caused billions of dollars of damage to corporate networks, but new types of malware have been created that specifically target laptops and handheld devices. Moreover, malware creators now use short-span attacks to transmit the malware to millions of users in a few hours; serial variant attacks, whereby variations in the malware allow it to subsequently evade detection; and designer malware targeting specific targets so there is no chance to identify the malware and develop a signature of the attack. A number of specific malware programs have been created to shut down cell phone and other handheld devices, and, worse still, roaming laptops can acquire malware and then infect corporate networks when users return to the office and attach to the corporate LAN from inside the firewall. A common way for a corporate network to become infected is when users synchronize home and work computers, inadvertently transporting malware. Adding to this problem is that most updates for spyware are only linked to corporate networks, leaving mobile devices vulnerable. The mobile blind spot is a problem created by the fact that mobile devices of teleworkers can travel for weeks without direct contact with the network, and thus miss updates of security in the interim.

Phishing or social engineering involves efforts to dupe computer users into sending confidential information to a third party. Sending bogus e-mails, online contests, persuading persons to download items with Trojan horses attached, and drive-by downloads are all ways in which phishers can obtain and exploit personal data. Phishing can become so devious that even if a user visits a legitimate site he or she may inadvertently download a secretly implanted malware, or in the case of MySpace overlays of redirecting URLs were utilized to download malware into victim PCs. As many as one in ten of all the URLs on the web will attempt to perform a malicious act against site visitors.

Mobile users are also described as veritable "sitting ducks" when it comes to direct hacking. In terms of spoofing, wireless communication is vulnerable because of how easy it is to intercept it, and because they often connect to hotspots that can in turn be spoofed by hackers. Most Wi-Fi connections do not have encryption capability, with the result that data are literally flying through the air, easily captured by those with malicious intent. A sniffing tool called Wireshark can sniff out the presence of mobile workers using a Wi-Fi hotspot to e-mail the office and thus gain access to corporate networks. Also, again because Wi-Fi hotspots offer no encryption, any usernames and passwords can be sniffed and stolen. Many mobile computer users from remote locations cannot distinguish between real and spoofed hotspots, meaning that while logging on they are giving away personal information.

The amount of loss and theft of mobile devices and the resulting loss of personal information is rampant. A clearinghouse that records these crimes suggests that the number of personal records exposed since the beginning of 2005 now exceeds 500 million. The fact that the increasing number of teleworkers are also downloading data into USB thumb drives and handheld devices has contributed to this vulnerability. There are many cases where IT managers have found company employees were using USB storage devices, despite a clear corporate policy stating that anyone found storing data on removable devices was subject to termination. Security is still a more difficult problem for many companies, who usually cannot afford IT professionals to advise them in security matters. In general, the approach to security in telework and mobile computing places the burden on management.

## Conclusion

Do people overplay the risk of cyberwar and cyberterrorism? They are sexy, and they do get media attention. And at the same time, however, people underplay the security risks of cybercrime and the risk of the telework and mobile workforce. It is impossible to eliminate all security risks to teleworkers and mobile computing that drive cybercrime. However, confronting and addressing these risks and concerns could reduce the frequency and magnitude of future incidents.

As telework and mobile computing increase, organizations must act to reduce vulnerabilities before criminals can exploit teleworkers and mobile computing. No single organization or program alone will reduce cybercrime, but combining efforts and aggregated initiatives can be effective. Telework and mobile computing information security risks and concerns are real to due to the sheer number of teleworkers and rapidly changing information technology. The greater dependence on communications requires increased remote access to an organization's data to perform the work and is to some extent a function of the increasing geographical distance among workers and between workers and the headquarters. The increase in the number of teleworkers and the frequency and scope of official and unofficial telework has exposed a vast amount of confidential organization data outside the physical office and has increased the likelihood of security breaches. The ever-growing use of teleworking increases convenience and efficiency; however, the mobile workforce can have significant implications on organizations' data security and information technology operational strategies. Even though organizations believe in a technological solution to this problem, there is another solution that is often times overlooked, namely: enhancement of mobile workforce security through peer pressure on teleworkers' adoption of information security measures.

## Which Factors Motivate Teleworkers to Take Necessary Security Precautions?

The Security Journal, Summer 2009

Telework--the practice of using off-site or portable computers to perform company or agency-related duties--has become increasingly commonplace during this first decade of the 21st century. This ever-growing use of telework is convenient and popular with workers but the jury is still out on whether it increases productivity. Telework arrangements can, however, have significant implications on an organization's data security and information technology operational strategies. A greater reliance on telework raises numerous information security concerns, including everything from breaches of confidentiality, increased opportunities for unauthorized viewing of data, data theft, and data leakage. Is there a solution? As the need for flexible work arrangements continue to be an important factor for workforce management, this article will highlight for management to have clearer understanding of how security policy are received, perceived by their end users, and the use of technology for organization data protection.

Telework is undoubtedly on the rise, especially for Government agencies. Because it is a mandated program, Federal employees are almost three times as likely as private sector employees to have an option of working from home. Until recently many Federal teleworkers had no choice but to use their own computers and equipment simply because some Government agencies either could not afford to supply their teleworkers with secure, Government-issued models or were not willing to take on the expense or complications involved. However, it is becoming increasingly more evident that the price of the equipment is minimal compared with the cost of dangerous security leaks that occur as a result of using non-secure, unauthorized technology.

To address the problems related to telework data security, some key terms must be adequately defined. In general, teleworkers are employees who work outside of the official workplace by way of a technological connection during regularly scheduled work hours—at home or an alternative workplace—on full-time, part-time, or situational basis. Non-teleworkers are considered employees working at the official workplace during regularly scheduled work hours. An unofficial teleworker is an employee who works at an official workplace, yet also performs work-related duties off-site on nights or on weekends.

Teleworkers use various client devices, such as desktop and laptop computers, cell phones, and personal digital assistants (PDAs) to read and send e-mail, access Web sites, review and edit documents, or perform other tasks. Most teleworkers use some type of remote access to an organization's resources and data from locations other than its facilities. In many instances, teleworking has become an extension of mobile working, rather than being simply one or a few workers based outside the organizational perimeter and accessing the network from time to time. Organizations now have both teleworkers and mobile workers and require support for information technology devices.

The ability to be productive outside traditional office hours means that work can be extended in the dimension of time, and the fact that it is possible to work outside the office means that work can also be extended in the dimension of space. As the temporal and spatial territories of office work are being extended, so too are the boundaries of the organization. Certain items of technology, which we call work extending technology (WET), facilitate expansion of an organization's borders. Mobile phones, laptops, and BlackBerry devices provided to staff make working anywhere at any time quite simple, thereby affecting the traditional boundaries between two separate spheres or experiential categories identified as "home" and "work."

Telework has a critical attribute that differentiates it from other types of work organization, and of portable information and communication technologies allows it to be "delocalized." In other words, work is free from the confines of a particular physical work environment. The growth of the new technologies has created a distinct phenomenon: when a corporate identity requires only an e-mail address and a Web site, there is no need to be physically based in any particular location. In the course of a day, work can be conducted in a home, in a car, at a client's office, in a restaurant or café, walking on a street, on a park bench—literally anywhere. The convergence of the computer, television, and telephone allows these communication tools to offer increasingly broad coverage.

The social meanings of home and work environments are beginning to merge as they become informationally similar. Convergence, a seemingly benign term, entails a whole host of technological dependencies that affect our daily lives. In the context of telework it implies an increasing lack of differentiation between our work lives and our domestic lives.

## Challenges

Teleworkers--official and unofficial--present unique challenges for an organization due to the information technology needed to provide them with a secure working environment while implementing security controls. Yet despite the ongoing information security issues related to telework, as shown in the chart below, the purported cost- and time-savings are making it an increasingly popular choice in Government agencies.

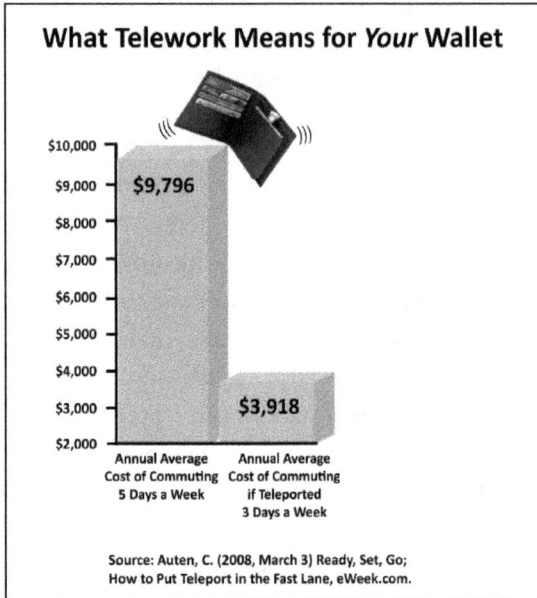

## What Telework Means for *Your* Wallet

$9,796 — Annual Average Cost of Commuting 5 Days a Week

$3,918 — Annual Average Cost of Commuting if Teleported 3 Days a Week

Source: Auten, C. (2008, March 3) Ready, Set, Go; How to Put Teleport in the Fast Lane, eWeek.com.

Chart 1

In fact, there is currently an attempt to make telework an even more significant part of the Federal workforce, by introducing H.R. 4106, which is the Telework Improvement Acts of 2008 bill and if enacted would mandate the option of a telework arrangement unless an agency can prove that such an option was not viable. H.R 4106 would also require that Federal agencies incorporate telework into the continuity of operations planning and for mission critical personnel to be equipped to telework in time of a catastrophe.

There are several reasons why the Government is pushing for more teleworkers. Telework offers substantial savings of physical facility related costs including rent, storage, electricity, including decreased energy consumption and less traffic (resulting from eliminating the need to drive back and forth to work; see Chart 1, above), as well as a greater level of personal contentment for employees who are not forced to deal with the stresses of office life. This is, of course, is expected to result in greater productivity.

Opportunities for productivity improvements as workers no longer suffer from the tiredness associated with physical commuting and are removed from many of the distraction of a traditional office. Telework has our interest in the potential for a dramatic shift in the workplace and work process based on evolving telecommunication technology--the history of the world is the history of change.

Telework has caused the line between personal time and work time to become blurred in such a way that it seems natural for a person to take care of personal business at a time when the person should be taking care of work business. Telework brings the public domain (work) into the private domain (home). This raises a number of issues for teleworkers, for example, the importance of separating being at work from the family time. This, of course, results in a decrease in productivity and also raises the problem of requiring a workforce that is highly self-motivated – something that is often easier said than done.

As a theoretical construct, telework works well. However, a problem in its application exists in that expansion of teleworking employment arrangements can have significant implications on an organization's data security and information technology operational strategies. Increased reliance on telework increases risks for information security.

Security threats in a telework infrastructure are often related to the computers literacy of the teleworker accessing the network rather than the actual corporate network. Or computer-savvy teleworkers who link an organization's laptop to high-speed cellphone networks simply to avoid paying for special cards and service plans because they are able to easily access Wi-Fi hotspots when at home or traveling. In addition there is that unknown activity by employees at home, when information can be damaged, copied, or simply assessed in an unauthorized manner. And unethical employees can engage in activity against the business's best interests.

Information technology departments do have less control and oversight of remote users. Most organizations have facilities with physical control safeguards in place for onsite employees. Other concerns, who is also in the teleworkers home and what are their intentions, shoulder surfing watching entered IDs and password? Who else has access to the company equipment and the data being used, do children play on the equipment and able to access the organizations network?

With the increased benefits afforded by teleworking come increased information security risks. When taking into account the negative aspects of information security concerns and combining them with the questionable benefits of increased productivity, a question can be posed about why telework is on such a steady rise. When a teleworker combines personal business and office business on the same computer, not only do lines between personal and business matters blur, but more importantly, information security can be more easily compromised. A sophisticated intruder may find it easier to attack the laptop of a teleworker logged on to the corporation network than attacking the corporation network itself. The results very costly and damaging to the reputation of the organization.

With the lines between work and home becoming increasingly blurred, some of the most challenging questions faced by organizations relate to the difficulty in identifying and assessing the continued information security risk and the need to develop and implement effective information technology security controls to secure the data of employees who work from home with permission and without permission.

## Unofficial Teleworker

Of course, with every argument, there is a counter-argument, and not all of the information security concerns are that teleworkers cause a greater threat to security. More than half of U.S. Government employees unofficially work at home on nights or weekends, raising concerns about the security of the data on which they are working.

According a survey by the Telework Exchange, a group that advocates for more telecommuting opportunities, 58 percent of Government employees work from home without permission. In addition, 54 percent of those unofficial teleworkers carry files home. Organizations can do everything to protect information and data, but the unofficial teleworkers create a risk when working from home on unsecured computers and mobile devices.

Conversely, some reports proclaim that Federal teleworkers are actually less of a security threat than traditional office Federal workers. The report comes from the Telework Exchange and explains that the reasons security threats are reduced by way of telework is that materials are not being physically transferred from place to place and that teleworkers tend to be monitored more strictly than in-office workers.

The Telework Exchange survey results are from "an online poll of 258 Federal employees including sanctioned teleworkers, non-teleworkers and non-teleworkers who unofficially work at home revealed that Federal data is significantly more mobile and still vulnerable." The questions asked in the poll varied, but the most significant questions garnered the following results:

> "The report found that 63 percent of respondents who worked from home unauthorized—more than half of the non-teleworkers surveyed—used their home computers in doing that work. "People were saving documents on their home computers that were unprotected," said Josh Wolfe of Utimaco, a data security company that underwrote the study...When teleworkers and nonteleworkers where asked if they had antivirus protection on their laptop or desktop computers, 94 percent of teleworkers responded yes, while only 75 percent of non-teleworkers said yes (Sternstein, 2007)."

On the other side, 90 percent of workers in traditional offices surf non-work-related Web sites during working hours, causing additional demand on bandwidth and infrastructure. It is reasonable to assume that workers in remote locations do the same at least as often. At the traditional work site, workers are aware that they are using the company connection to make personal purchases, write e-mails, visit chat rooms, play games, conduct personal business, or generally wander around. It is clear that such activity is often monitored or, if not actually monitored, can be monitored quite simply. However, at home, it is less clear that one is using company resources for private activities on company time. The distinction between company time and personal time, company resources and personal resources, is unclear. Availability and use of monitoring devices in workers' homes by way of the Internet comes dangerously close to wiretapping, which is illegal under most circumstances. And, such monitoring is certainly an invasion of the privacy of one's home.

### Two sides

Information security and privacy are two sides of the same coin when it comes to the dangers of telework. The information security side comes from employers whose responsibility is to keep their customer and corporate data away from the prying eyes of competitors and others whose interests are a threat to the organization. The privacy side comes from employees who are trusted to handle the data and to keep them safe, yet are at risk of losing their own privacy when this effort takes place outside the physical office environment. There may be an invasion of privacy associated with surveillance of the teleworker by the organization but there are also technologies that will actually conduct surveillance, so the potential to use them does exist. Even if an occasional visit to the home may place the household under surveillance, this could in a way constitute an invasion of privacy of the entire household.

Information security and privacy in cyberspace can be especially difficult, requiring elaborate firewalls and multiple checkpoints that seem to be violable nearly at will by sophisticated hackers. A more complex problem arises when organizations implement powerful security systems and discover that they have a negative impact on the functionality of the software that allows them to do business in the first place. Effective information security devices can produce decreases in application response times, causing significant delays between data entry and display, limiting call completion when using Voice of Internet Protocol applications, and creating a generally sluggish electronic work environment. In addition, the growth of the remote electronic storage industry, already strong in the late 1990s and fueled dramatically by the terrorist attacks of 2001, has become a costly and unwieldy business requirement.

## Motivations

It is vital for organizations to make efforts to convey to teleworkers that data protection and information security is important to an organization and employees action make a difference in achieving the overall goal of protection of sensitive data. Despite increased awareness and training on security issues, many employees do not take the necessary precautions for deterring security risks. Knowledge and training are useless if not implemented and not perceived as important. So, which factors motivate teleworkers to take the security precautions that will help them avoid the mishandling sensitive data? Although, security policies and security policy compliance is very important aspect of organizational information security, the perceptions about the severity of breach and response efficacy are likely to affect teleworkers compliance intentions. Harmonization between management and teleworker perceptions about organization data protection and information security values plays a role in teleworkers data protection and information security behaviors. There is a greater security threat when the teleworker and organization do not have alliance regarding data protection. Teleworker perceptions of the importance of data protection is relevant with the policy compliance behavior along with the role training and awareness and policy enforcement play in shaping the information security climate perceptions of teleworkers. Policy compliance intentions are predicted by the management information security importance perception, which in turn is highly associated with the employee perceived training and awareness as well as policy enforcement efforts by management.

Generally, people intend to behave a certain way when they evaluate it positively, when they experience social pressure to perform it, and when they believe that they have the means and opportunities to do so. Those factors help the motivations of teleworkers to comply with information security measures as well as the factors that may inhibit their compliance. Information security breaches in organizations have the potential for widespread damaging consequences for a long list of stakeholders of any organization. Understanding what motivates teleworkers and managers for taking data protection and information security problem seriously will assist and encourage them to devise and implement effective solutions.

## Consideration

Telework has caused the line between personal time and work time to become so blurred but more importantly, information security can be more easily compromised. The need of the Internet is so important for teleworkers to get work done and there is not accountability for the bad guys on the Internet. When using one computer it has become natural for a person to combine personal business and office business. There is greater concern when the teleworker travels with one computer. This is where line between personal and business matter becomes even more muddied.

For both today's and future generations, telework will become even more blurred with lots of work, and lots of personal business, requiring more access of things on the Internet. A viable solution would be to give the teleworker more flexibility by the use of a technology that splits the difference between more accountability of the organization data and a computer that teleworkers could use more freely. Create a Red-Green environment for the teleworker in which to do both personal and work business. Such a virtual solution is just one of many solutions of trying to physically separate different use contexts would use virtualization on the desktop to logically separate context. The 'virtual solution' would allow users to separate work from play without having to use physically separate devices. Computers and laptop could be designed to pretend to be more than one machine to satisfy the desire for information security without a full lockdown. In its simplest implementation a device could be divided into two virtual machines: "Red" and "Green." The point is to partition the device into two worlds; "Red", less safe and unaccountable and "Green", safe environment and accountable.

## Conclusion

The importance of telecommuting and the information security challenges it raises will grow alongside globalization, and the need for flexible work arrangements will continue to be important factors in workforce management. Information security concerns for telecommuting are important because of the sheer number of teleworkers and the rapidly changing information technology. Our greater dependence on communications requires increased remote access to an organization's data to perform the work and is to some extent a function of the increasing geographical distance among workers and between workers and the headquarters. While it is important to have information security procedures such as training and awareness initiatives and policy enforcement mechanisms; it is also vital that management have a clearer understanding of how these messages are received, perceived by their end users, and the use of technology for organization data protection.

## Data Security and Privacy Issues in the Digital Era

The Security Journal, Fall 2011

*Good name in man and woman, dear my lord, Is the immediate jewel of their souls; Who steals my purse steals trash; 'tis something, nothing; Twas mine, 'tis his, and has been slave to thousands. But he that filches from me my good name Robs me of that which not enriches him And makes me poor indeed.*
*—William Shakespeare's Othello*

### Introduction

Pick up a newspaper or turn on the television and there are reports of massive data breaches (e.g., WikiLeaks). With the increased use of technology for everyday activities from ordering a moving online, conducting financial business, and updating Facebook, the threat of data security in the digital era has become a serious issue. For example, when the Internet was created using 1970s technology, it was not intended to be a global infrastructure used by billions of people. The security of the Internet had not advanced to match its capabilities. As stated in the report of the Commission on Cybersecurity for the 44th Presidency, "Cybersecurity Two Years Later," the problem with the current state of data security and privacy is that:

"The creators of the Internet believed cyberspace would become a self-organizing global community led by private action, where governments should play only a limited role. It would become, in this rosy view, a global commons where people could invent and create without constraint. The problem is that the lack of constraints empowers malicious activity as much or more as innovation. Other governments increasingly reject this pioneering American vision as inadequate for securing what has become a critical global infrastructure."

## Technology, Data, and Who Gets Access to It

The advent of the digital era with the ubiquitous use of smartphones, wireless technology, and the flow of sensitive data between computer systems has many wondering if true privacy is possible. Cameras and computers track the activity of people, both in public and private places. Cameras capture images that are stored on computers without the subject ever knowing about it. Steve Lohr notes in a recent article, "High-resolution, low-cost cameras are proliferating, found in products like smartphones and laptop computers."

Other technologies allows the user to watch others using programs such as Google's Street Wise or to take pictures and then mine the internet for matching images and other personal information using Goggles. The use of "cookies" and other tracking devices and programs enable data to be collected without the user's knowledge or permission. Every time, an individual uses his or her smartphone to make a call, text, or search the Internet, the company providing the service can collect location and other information. Once information has been collected and stored it becomes available to anyone, human or machine, that can gain access to the data.

Increasingly, people have lost the ability to determine who has their information, how it is used, and how securely it is protected against cybercrime. In the digital era, electronic communications systems generate vast quantities of transactional data that can be readily collected and analyzed. Unfortunately, the security against the illegal and unauthorized use of information and violations of privacy typically trail behind the hackers and other cyber criminals' ability to engage in these activities.

Technology is predicted to advance to the point where computers not only will gather information but also engage in the analysis of the meaning of those images. According to the Lohr (2011), the U.S. government is working on developing a program called "The Mind Eye" that would be able to gather information, analyze it, and report its findings. While these types of technology can be extremely beneficial, they also can be extremely dangerous if used for nefarious purposes by hackers and other non-authorized entities.

With the introduction and increased use of cloud-based technology, current laws fail adequately to address data security and privacy issues of non-locally stored information. Those companies that deal with the storing of vast amounts of personal data such as Facebook, Microsoft, Google, and AT&T as well as privacy advocacy groups such as the Electronic Frontier Foundation are pushing Congress to update the 1986 Electronic Communications Privacy Act (ECPA) to include the warrant requirement by all agencies. The current ECPA, addressed wireless voice communications and electronic communications of a non-voice nature, such as email or other computer-to-computer transmissions.

Another concern is over the collection and use of data by legitimate government and law enforcement agencies. Berman & Bruening (2007) noted technology has freed law enforcement intercepts from the constraints of geography, allowing intercepted communications to be transported hundreds or thousands of miles to a monitoring facility. In June, 2011, Senator Patrick Leahy, chairman of the Senate Judiciary Committee moved forward to update digital privacy laws that are 25 years old and really do not address issues of data security and privacy related to the Internet, smartphones, and other similar technology.

In May, 2011, Leahy introduced the Electronic Communications Privacy Act Amendments Act of 2011 that would require police and other government agencies to obtain a search warrant to access private communications and the locations of mobile devices. However, bowing to pressure from law enforcement on data collection and use restriction, there would be no restrictions on doing a search and obtaining information from the Internet or from recording an individual using the GPS from a smartphone. According to McCullagh (2011), this is part of the problem in creating laws and policies used to protect data security and privacy. There is always going to be some governmental or even private entity that will make the case for why it should not be subject to the restriction on data use and violation of privacy.

### Legislation Enacted to Protect Data Use and Privacy

On the other hand, over the last fifty years, there have been increased efforts to limit the extent that law enforcement and other agencies can use electronic surveillance. In 1967, the Supreme Court has limited electronic surveillance to those who obtain a warrant based on from a judge with a justifiable need to engage in the surveillance and limits of the information that could be collected. In 1972, concern over the protection of data collected and stored on computers resulted in the Privacy Act of 1974 that contained in it a code of fair information practices, which has served as the foundation for subsequent privacy legislation. In 1996, the Health Insurance Portability and Accountability Act became law and provide greater security and protection of patient rights in the collection and use of medical and health information. The Children's Online Privacy Protection Act was enacted in 2000 and provides protection of a child's personal information that is collected from commercial websites by requiring parental consent before information can be collected. In addition, during the 1990s greater efforts were being made to protect the collection and use of data business obtained about consumers conducting online transactions. Unfortunately, this is an area where data security and privacy lag in adherence to fair information practices.

### New Conception of Privacy

As cited by Breman & Bruending (2007), in the digital era, the concept of privacy is changing from that defined by Justice Brandeis in 1898 as "The right to be let alone." The ability to collect, store, manipulate, and transmit vast amounts of data anywhere in the world in milliseconds has altered this conception of privacy. The concept of privacy is changing with the advances in technology. The idea now is not so much keeping aspects of one's private life "private" but of limiting the number of entities that would have access to this data. Therefore, data security becomes extremely important. The issue is now autonomy and control over one's personal information. Stated by Breman & Bruening (2007), in the digital era, privacy entails an individual's right to control the collection and use of his or her information, even after they discloses it to others.

While absolute privacy is highly improbably in the digital era, what people want is, for their information be used for the purposes for which it was provided (e.g., conduct a financial transaction, get medical treatment, purchase a product or service) and have the power to prevent other uses of that data without permission from the individual. Under the principle of fair information practice, people must be notified when their information is being collected, used or exchanged. They must be able to have control over their personal data use and be able to have access to this data when requested. Most important, people must be assured that their personal information is accurate and kept secure from cybercriminals as well as government intrusion and illegal surveillance.

### Efforts to Protect Privacy and Secure Information

It is difficult to protect data that is transmitted through the Internet. This is because the open decentralized architecture results in data being transmitted through different paths over many computer systems. This makes the data, particularly susceptible to being intercepted. In order to help users become better informed about information security and use policies on the Internet, Platform for Privacy Preferences technology was developed by the World Wide Web Consortium. Websites with P3P technology inform the user about the site's privacy policies and provide the user with a level of control and choice over whether once informed the user would want to continue on that site.

The individual does have some tools available to provide a level of security and privacy. New applications provide for engaging in anonymous or encrypted transactions. According to Berman and Bruening:

> "Some tools developed to protect privacy exploit the decentralized and open nature of the Internet. These tools may limit the disclosure of information likely to reveal identity, or decouple this identity from other information. Others create cash like payment mechanisms that provide anonymity to individual users, vastly reducing the need to collect and reveal identity information."

In the report of the Commission on Cybersecurity for the 44th Presidency, it was recommended that a National Strategy for Trusted Identities in Cyberspace (NSTIC) be developed and implemented in cooperation with the National Program Office in the Department of Commerce. The goal is to create stricter standards for authentication of identity and to encourage those in business (especially e-commerce) and individual users to make use of online authentication procedures. However, the report states, a continuing problem has been the choice by many people and businesses not to use the existing authentication technology that is effective in increasing data security and privacy.

The most powerful force to get lawmakers and others to increase cyber security is the general public that becomes engaged and takes action (e.g., using products from companies and voting for politicians who are strong advocates of protection and privacy laws). The collective voice of technology users can cause a change. When individuals mount a collective protest and put their words to action, they can force government and private sector business to provide greater protection. For example, DoubleClick had planned to use technology to link information collected offline and online. When this outrage was voiced, the negative press resulted in DoubleClick's stock prices to fall and the company to abandon its plan. In another example, Intel's Pentium III microprocessors were enabled with the capability of tracking user activity across the World Wide Web. The collective voice against Intel resulted in the company discontinuing this processor and disabling the tracking ability on microprocessors already installed.

### Conclusion

Is data security and privacy possible in the digital era? Yes. However, the United States and other countries must rethink its policies towards cybersecurity in an environment where capacity, speed, and connectivity advance at increasing rates. In the future, new technologies, increased dependences on worldwide communications, and living as a global village will create more data security and privacy problems. There is a need of a framework, a way to conceptualizing privacy that provides a clear understanding of privacy across boundaries. In the last two years, there has been a renewed commitment by the United States to develop more effective policies and legislation to provide a level of protection to match the technology used. Developing effective and comprehensive nation strategy for cybersecurity both within the military and civilian sectors has become a priority. This includes engaging other nations to work cooperatively to develop cybersecurity, even among those nations who want to end the United States' perceived hegemony over the Internet and to reshape the Internet to serve their national interests.

## Information Security against Theft and Disruption

The Security Journal, Summer 2009

### Introduction

Network security is an incredibly important issue, but the sheer importance often leads to an oversight: physical security. While millions are spent thwarting hackers and identity thefts, for example, much crucial data is no more physically secure than a laptop computer left in an unlocked car. Data is so easily copied that sensitive information can be purposefully of accidentally emailed or otherwise sent to someone –for solid, business-related purposes – whose own desktop or other data storage devices may lack the security necessary to protect crucial data from exposure.

Cyberattacks may be geared toward disruption or denial of service, or financial gain. Financial gain is sometimes associated with identity theft or reallocation of funds from accounts, both of which are considered white-collar crimes and do not receive much attention from law enforcement agencies or the general public. An example of financial loss was demonstrated in a survey of 611 companies doing business on the Internet in 2001. Results indicated that 83% of the companies experienced security breaches and 62% said there was some type of financial loss, with average damages of $47.8 million, compared to $26.6 million in 1999. These were the relatively early days of information security concerns, and the height of the dot.com boom. Surely the situation has improved since these days, now that the Internet is more fully integrated into economic life? Unfortunately, no.

Information security failures are still a major economic fault line.

The costs of poor information security has increased since the 2001 dot.com boom: "In the 2006 CSI/FBI Computer Crime and Security Survey conducted by the Computer Security Institute and the Federal Bureau of Investigation, 313 computer security professionals reported a total of $52.49 million in losses linked to computer security incidents for 2006." This, despite the immense amount of money spent on computer security in the past decade and even the emergence of dedicated information technology security divisions within larger firms, as well as many number of vendor firms working as providers of various security protocols, training, and hardware/data management services.

Much of information security failure activity comes in the form of illegal access into a company or government network system. Some hackers start out as outsiders who use various password attacks in order to gain access as a user of the network. Many begin as insiders, in the form of disgruntled or former employees. They will use known weaknesses in the system to gain further access as an administrator or super user. Once these privileges are attained, a hacker then can read or alter files, control a system, and insert a rogue code such as a virus to damage the infrastructure. Often, hackers may not even be motivated by gain or aware of what systems their viruses have infiltrated.

The costs of security breaches can be severe, and evolve into not only crime but literal "cyberterrorism." There are several possible scenarios for cyberterrorism. In one, a cyberterrorist hacks into the processing control system of a cereal manufacturer and changes the levels of iron supplement. Customers suffer, some perhaps die, the product is recalled, and the firm suffers greatly. In another, a cyberterrorist attacks the next generation of air traffic control systems. Two large civilian aircraft collide. In a third, a cyberterrorist disrupts banks, international financial transactions, and stock exchanges. Economic systems grind to a halt, the public loses confidence, and destabilization is achieved.

As technology continues to evolve, so will opportunities for cyberterrorists and other malevolent actors whose goals may be financial, simply apolitical, or utterly inexplicable. The phenomenon of "cybercash", in which nearly all money is electronic, and thus instantly transferable and exchangeable, is in the offing, and as Guttmann (2003) notes:

> Given the inherently unstable nature of self-regulation, cybercash will need external stabilizers for support. Facing a heterogeneous money from appearing in many variations, such stabilizers must contain central-control mechanisms which manage the organizational complexity of this new monetary regime. Those mechanisms have to ensure a systemic coherence and robustness of the electronic-money regime in the face of possible technological mishaps, cyberterrorism, incidences of financial crisis, and our economic system's propensity to monetary instability.

Given the stakes, both for society as whole and individual firms, the security of information and of the information infrastructure are paramount. The very nature of information —it can only be shared via copying, for example, and not exchanged —limits its total security, and the fact that information must be used in order to have value means that information will be exposed to end users when transmitted. Both external loci (hackers, cyberterrorists), and internal loci (employees, vendors, physical objects such as hard drives and servers) for security vulnerabilities.

### Risk Assessment

Risk assessment involves answering these seven questions:

1. What can go wrong? (threat events)

2. If it happened, how bad could it be? (single-loss exposure value)

3. How often might it happen? (frequency)

4. How sure are the answers to the first three questions? (uncertainty)

5. What can be done to remove, mitigate, or transfer risk? (safeguards and controls)

6. How much will it cost? (safeguard and control costs)

7. How efficient is it? (cost/benefit, or return on investment analysis)

There are six major steps involved in a risk assessment:

1. Inventory, definition, and requirements
2. Vulnerability and threat assessment
3. Evaluation of controls
4. Analysis, decision, and documentation
5. Communication
6. Monitoring

These steps are of course part of any sort of risk assessment and are not specific to information security. This paper will examine specifics of how a risk assessment for information security may look and how this could influence top management views on information security management.

### Inventory —Threat Assessment

Bielski (2004) quoting a compliance/ policy management official explains the first step of developing and information risk management program:

"Overall, you need to know what your hardware, operating system, and application layer security vulnerabilities are and you need to know, generally, who uses what application and why. Then, you need to lock down these assets in a systematic way. This is simple but it isn't easy."

Larger organizations may not even know what they have. If one very large institution could not produce a network map in a reasonable time-that meant, they had servers and other digital assets that were unaccounted for and reliance on default controls at the hardware and operating system level and inconsistencies in response time to threats that can be dangerously complacent. Data are notoriously difficult to "rein in" once made available on certain networks. Data often end up stored on local files rather than being centralized, and once out of control of those in charge of it, data can end up virtually anywhere in the world. Not only is the Information Technology Department an important part of this initial assessment, so too is the Human Resource Department.

Indeed, risk assessment begins not necessarily with risk alone, but with a complete assessment of information technology capabilities and vulnerabilities. Often organizations fix point problems without looking at the overall picture. Large companies are the most notorious for doing this, having separate business units that tend to fight more often than work together for the good of the company. Subsequently, without cooperation and an overarching plan, they end up vulnerable to attack. An initial assessment will also allow the firm to know where its data are, and what possible damage can be done via unauthorized access to these data. Clearly, a firm with trade secrets or even sales leads can be severely compromised by a disgruntled employee or a former executive taking information to a competitor. Firms with access to consumer retail information (e.g., credit card numbers, social security numbers, and other such information) may also be open to significant liability if proper protocols and industry standard protections are not followed. Staffers may need training on security protocols and even the basics of common hacker tactics (e.g., viruses, social engineering, etc.) However, just knowing these possibilities is not enough — a top-to-bottom security assessment must be put into place in order to understand what an organization's specific vulnerabilities are, and how exposure can be limited.

Subject-matter experts, including Certified Public Accountants, can also be brought in to deal with the non-technical ends of security issues. According a Certified Public Accountant understand internal controls and can assist companies in adapting control concepts to protect the confidentiality, availability, and integrity of valuable assets. Certified Public Accountants can perform trend analysis and search for patterns in security data. They can be assets in drafting policies and procedures for safeguarding financial and customer data and also play a role in training employees to abide by the rules. Certified Public Accountants can be useful as they often have interpersonal training that technical staffers and IT professionals lack, and have a greater professional understanding of data control and non-technical protocols in general.

In addition to the application/user layer, the operating system itself can often have vulnerabilities, especially as operating systems these days are often tied to specific applications, such as web browsers, or even databases such as the .mac storage facility that comes along with recent versions of some Mac operating systems. Operating systems can be more or less vulnerable to viruses, Trojans (which do not attack a system, but simply draw information from the system), and other malevolent programs. Windows-based operating systems are especially vulnerable to such programs due partially to the widespread deployment of the systems (thus, more vectors from which to spread) and due to animosity toward Microsoft as a firm in the hacker community.

Hardware is also vulnerable. Many firms "give" employees laptops, pagers, palmtops and other easily portable computer equipment, which can be stolen, destroyed, or delivered to an opponent for hard drive copying after hours. Old equipment can be disposed of without the data being erased from it — indeed it can be said that data never truly "goes away"; with sufficient resources virtually any hard drive or other information medium can be milked of "deleted" information if the equipment remains intact. Recommend contracting with asset disposal firms to ensure that old hardware does not end up on eBay, with information intact, or in the hands of a competitor, criminal, or terrorist who will often need little more than a Norton utility restoring program to retrieve "erased" data.

### Analysis — Monitoring

Unfortunately, the demands of customer service and the realities of e-commerce more or less demand a level of decentralization. Information is valueless if it is not accessible, and regulations are not useful if compliance levels are low. For example, perhaps the best way to protect passwords would be to demand daily changes — such a protocol would be useless, however, as compliance would swiftly drop to near-zero. Firms need to determine what level of employee access to give and how much employees should be generally trusted. To the extent that you can set policy, automate it, and forget it, you're better off.

The desktop is perhaps the symbol of the decentralized information infrastructure. Desktops can theoretically, if connected to the network, give any employee access to all the data in the firm, and almost certainly too all the email addresses in a firm. All it takes is a virus in the right address book to compromise an entire system. Effective security begins with perimeter-based defenses like intrusion detection, patches, and anti-virus protection and continues behind the firewall, so that data whether 'at rest or in motion' is encrypted when not under authorized use. However, past the perimeter sits the employees. Lineberry (2007) describes a security test based on exploiting social engineering vulnerabilities:

> "The tester entered bogus credit card information with his order, which triggered a phone call from an employee whose job it was to manually enter all payment information from the seller's online orders. The tester offered to send the correct information to the business in an e-mail with an attached document. The phone conversation was designed to lower the guard of the employee, who might not have opened the e-mail had it come from a complete stranger. In the end, the employee opened the e-mail and attachment and welcomed in the Trojan, allowing the tester to break through the firewall and gain access to 25,000 credit card accounts."

As we can see, information security does not end at the desktop. It also must be embedded in the scripts and protocols given to call center staff, tied into physical plant security, and even connected to hardware and waste disposal. Recommend that any information security risk assessment involve answers to these questions:

1.      Are employees educated and aware of common information security threats?

2.      Do they write down or freely share passwords with others?

3.      Do visitors freely move about facilities without facing barriers to entry, such as a requirement to wear a company-issued badge?

4.     Is it common to see sensitive information, such as completed employment applications or client documents containing Social Security numbers, accessible in unmonitored or otherwise unsecured areas?

5.     What is the prevailing employee attitude regarding information security controls?

6.     Are enforced information controls viewed primarily as a nuisance or a necessity?

Also a list of security breaches in order to dollars lost in 2006, which can assist in understanding where security efforts should be focused. The problems, in order, are:

1.     Virus contamination: $15.69 million

2.     Unauthorized access to information: $10.62 million

3.     Laptop or mobile hardware theft: $6.64 million

4.     Theft of proprietary information: $6.03 million

5.     Insider abuse of Internet access or e-mail: $1.85 million

6.     Bots (zombies) programs similar to viruses within the organization: $923,700

7.     System penetration by outsider: $758,000

8.     Phishing in which the organization was fraudulently represented as the
sender: $647,510

9.     Password sniffing: $161,210

The list is an interesting one as essentially undirected actions cost more than directed actions. Viruses are only rarely aimed at a single organization; instead they are designed to spread and the extent to which they infect hundreds or even thousands of systems is a testament to the ability of the virus writer. Downtime and even losses of back up information is the central cost of viruses. Unauthorized access is more often directed, though generally at the information per se (e.g., credit card numbers) —

organizations are targeted to retrieve this information and the most vulnerable of, for example, various online commerce firms, are simply the ones who end up being the victims of the attempts. Only further down the list, but still very important and expensive, are directed attacks against organizations in order to access their information. Insider abuse was ten times for expensive than password sniffing, suggesting that personnel training and employee morale are perhaps more important elements of a security plan than the deployment of anti-sniffing software, or frequent password changes.

### A Holistic Approach?

The answer to the thorny issue of information security and risk assessments for the same may be a "holistic approach", which brings together physical and information technology security. There are four methods to this: integrated identity infrastructure, centralized provisioning and de-provisioning, consolidated logging and auditing and the risk management approach. The balance of the article concentrates on the first method, integrated identity infrastructure.

Identity is crucial to security of all sorts —think photo IDs, fingerprint records, even surnames and street addresses as historical developments for the creation of coherent and unique identities. Businesses must audit their identity use to make sure that passwords or email addresses and other such codes aren't being "spread around"; all data must be accounted for, as must its physical locations. Then there must be a created standard of "truth" as far as determining the identity of users to avoid "social engineering" —hackers that contact some person in an office and ask for "help" recovering lost passwords or asking for information that under normal circumstances would not be shared, for example.

Some of the steps, such as "centralized provisioning", for

example, are fairly intuitive —one department provisions data for the entire organization, rather than allowing different departments, or even individuals, to handle their own data security (which would greatly increase risks all across the organization). One tool is virtual directory technology that allows the exposure of multiple directories and databases as a single directory is becoming a popular tool for this. Thus, information can be hidden or presented based on organizational needs and every instance of presentation or access to the directory can be recorded and monitored.

Some vendors are creating excellent applications for logging and auditing —"the user interface" of one, called Next Generation Management Infrastructure, by Checkpoint, is comprised of a security 'dashboard' which gives administrators a comprehensive view of all networked security devices, including third-party software. Administrators can determine whether servers are up and running or how many systems were scanned for viruses.

### Resources

Ultimately, the question of information security is tied to that of resources. Millions can be poured into training staffers not to hand out passwords or fall for the tricks of social engineers; contracts with vendors who will physically melt down hard drives to avoid unauthorized copying from legacy systems can be signed on a weekly basis; and everyone from the Chief Executive Officer to the part-time janitor can be trained in security protocols and quizzed for possible security vulnerabilities, but what if there are no barbarians at the gate after all? This is unlikely —viruses and Trojans needs not be directed specifically at any one computer network but instead gain power through propagation across many networks— but there does need to be a cost/benefit analysis put into place in order to determine how much security is enough, and how much is too much. There is a rule of thumb that larger organizations should have dedicated security teams. A general rule of thumb is that for every $1 million in revenue, approximately $2,000 should be spent on security. Another benchmark is 2 to 5 percent of the overall information technology (IT) budget.

Security expenses can be kept down by integrating security concerns throughout all business processes. The Human Resource Department should take computer knowledge, individual character, and other attributes into account during the hiring process, for example, and work with employees to keep them from turning disgruntled. Security is a special problem during transitions such as mergers, shifts in business processes, or layoffs, and so the Human Resource Department has a role to play throughout employee's working lives. The Information Technology Department, in addition to building firewalls, setting up password change programs and protocols, can assist by assuring interoperability of different systems to avoid expenses of having multiple security systems for various operating systems, servers, and other perhaps incompatible hardware/software combinations.

## Conclusion

Ultimately, security is an issue for top management, as only the executive suite can set the tone for information technology, human resource, the physical plant, and other departments in the firm. Security must be understood as more than a necessary evil — information security directly impacts the bottom line. More money exists today in "virtual" form and more personal information is stored in databases today than even existed several decades ago. The economy, in going global, has also gone networked. The modern enterprise must embrace the notion of a holistic approach to security bringing together physical and IT security. As the old saying goes, a chain is only as strong as its weakest link, so information security if it is not a priority for all organizations will increase the security risks for essentially any organization within a chain of value.

## Identity and Privacy in the Digital Era

The Security Journal, Summer 2011

*There is a growing consensus that if the jumble of state and federal statutes, consumer pressure, and self-helps to be unified into meaningful privacy protection in the digital age, then we will have to do more than pass a law. The law in general, and each of us in particular, will have to make some fundamental adjustments in the way we think of personal information and electronic communication. In doing so, we will ultimately have to change our idea of what we can reasonably expect to keep private.* —Ellen Alderman and Caroline Kennedy

### Introduction

In the digital era of the 21st century, when issues of privacy are discussed so are issues of identity. Closely related to privacy is personal identity: "when your identity is not known, you tend to have more privacy" (Cavoukian, 2006). When it is possible to link information to one's identity, this is considered personal information. The right to privacy is the right to exercise control over the collection, use, and disclosure of this personal information. Entities we trust to protect our personal information do not suffer when information gets exposed. It is individuals, who suffer when their personal information is exposed and do not have the capability to protect that information. Personal information is valuable and it becomes more valuable once stolen.

People are connected through their computers, mobile phones, the Internet, and a variety of other digitally based technologies. Each time an individual connects via digitally-based technologies; personal information is collected and stored, with or without that individual's knowledge or consent. Often this information is tracked and used by other by people, governmental agencies, organizations, insurance or marketing entities, and even thieves. Along with access to information, people also are losing personal privacy. Personal privacy and the protection of one's identity are serious concerns expressed by the general public and this concern over privacy is not new.

### The Nature of Privacy

For decades, the United States government has drafted legislation to protect personal identity and the exchange of personal information. The danger of the misuse of information was recognized nearly forty years ago when the U.S. Privacy Act of 1974 was established that included guidelines for basic privacy standards known as the Fair Information Practice Principles (FIPP). According to FIPP, those who use personal information must give notice to data subjects of their privacy policies. Data cannot be used or disclosed to others without data subjects' consent. Information stored must be protected from corruption or theft. Because we live in an era when transactions take place in without face-to-face interaction and identity is verified through the comparison of information recorded and stored in databases, FIPP requires that people should have access to the data that is collected and stored in order to verify data accuracy. Finally, policies and procedures must be established that enforce accountability of FIPP guidelines and outline punishment for data information use and security breaches.

Inherent in the FIPP guidelines is the idea that people should be informed by the data user when their personal information is being used. What becomes a central question is what yardstick is used to determine what is considered private and non-private information and what legally constitutes "fair" use of personal, private, and non-private information. Other questions include what is considered fair notice, consent, security, access, and accountability in the gathering, use, and exchange of this information. Essentially, the FIPP guidelines do not protect the actual data. It is written to provide regulation of the collection and use of the data in specific contexts.

### Regulatory Dilemma in the Digital Era

The explosion of digital technology and the ability to collect, store, and exchange vast amounts of information pose both enormous benefits as well as potential for significant harm. New storage technology has expanded the collection, organizing, and monitoring capacities of data users. Often the decisions about how the collected information will be used are done without human interaction; sophisticated pre-programmed algorithms automate the decision process on how the information is used and when and with whom this information is exchanged. Furthermore, there is virtually no cost to collecting and exchanging data over digital networks.

In addition to personal information, tracking technologies of digital information allows for other types of personal information to be collected and stored. This includes a digital trail of a people's behavior as they use different digital technologies. For example, the information that is collected by a grocery store when people use their club cards is used to develop a behavioral profile of peoples' spending behaviors. Digital technology provides capabilities of data information users to integrate and process information derived from multiple sources. This information is then reintegrated using state-of the art data processing software that allows raw data information to be dissected and recombined to develop digitally stored personal profiles of activity and behavior patterns. This information is then used to obtain insight about an individual that can be used in both positive and negative ways.

Digital technologies have provided public and private sector entities the capabilities to use information in-house and to sell or exchange this information across both public and private entities, often without the knowledge or permission of the data subject. The intrusion into privacy and has grown over the last twenty years as information technology grew increasingly sophisticated with the personal computer and distributed networks becoming ubiquitous. Now sole and small business owners along with large international corporations have the capabilities to gather and process vast amounts of data in whatever manner is desired. Public entities, often forced by shrinking budgets, increasingly have relied on private sector entities to manage the gathering and processing of information used by service providers. This has resulted in the privatization of public services. This trend has been increased with the use by state and federal agencies' reliance on private sector entities to use data collection and management activities to support antiterrorism efforts.

A problem arises in the protection of personal data because of differences in regulations that span across both United States state boarders as well as international boarders where different jurisdiction come into play as data is exchanged across these boarders. Those individuals in countries with strict data protection laws have concerns about the safety of their personal information when this information is sold or exchanged with an entity in a country with less stringent data protection legislation. One problem in this privacy issue is the different interpretations of privacy and what information can be collected and how it can be used. For example, is the searching of a computer hard drive through digital technology as one is connected to the Internet or network connection the same as a physical search of one's home? How does one know when this hard drive search has taken place?

What has happened is the explosion of digital technology has blurred the lines on what is private and public and how this information can be collected and used. This has contributed to the straining of the ability of privacy legislation to protect the individual from the collection and use of personal information. For example, in the name of national security, anti-terrorism legislation has expanded the power of state and federal agencies to gain access to information collected by private sector entities (e.g., e-mail records from private e-mail accounts, mobile phone records, and data from financial institutions).

### Privacy Rights

The challenge in the digital technology and privacy conundrum is how to provide a reasonable expectation of the protection of one's identity and privacy in the digital era such as fair levels of notice to individuals on what information is being collected and how it is being used as well as a self-regulation by those who collect and use this information. This protection of privacy is dependent on the policies, strategies, and technology that are developed and used by data users. In the United States, federal legislation has outlined privacy rights in the context of demands for a codified set of rules for how information is gathered and exchanged.

The challenge for data users is finding a useful set of formal guidelines that define the specific rights of people and organizations in relation to the ownership of personal information. To protect their interests, businesses want a consistent set of rules that are applicable across borders. Currently, data users must develop a coherent set of guidelines and rules based on the general FIPP principles that define the rights associated with control over personal information that is collected and used.

In terms of data privacy, the financial and medical industries are highly regulated in the use and exchange of personal information. The Financial Modernization Act of 1999 (Gramm-Leach-Bliley Act) requires financial institutions to notify their customers and given them the ability to "opt-out" of this information exchange when personal information is shared with affiliates or exchanged with third party institutions. However, while individuals must be notified about the sharing of their information, they typically do not have the ability to prevent information from being gathered or to correct misinformation collected. In the medical industries even more stringent regulations about the collection and use of information exists. Medical Privacy Rules enacted in 2001 requires written consent by individuals for their medical information to be shared and provides the right of individuals to review their medical records and have changes made if necessary.

Finally, different from the highly regulated financial and medical industries, many business sectors (e.g., direct marketing industry), have no formal regulations that govern the notice or consent of the use of personal information. Overall, businesses do not have to account for the data collection procedures, and people have no specific rights in this area.

### Monitoring and Enforcing Privacy Rights Regulations

When FIPP principles have been codified into information privacy protection regulations, these regulations only are effective to the extent that these specific rules are monitored and enforced. European countries are ahead of the United States in this area. Many European countries have established a comprehensive set of statutory rules that regulate how personal information is collected and exchanged by both government agencies and private organization. This allows for the enforcement of personal information use across all industries and sectors. This is different from the sectoral approach to regulation of the United States.

In many European countries, an oversight entity (often a national regulatory system) monitors compliance with a comprehensive set of regulations that govern how personal information is collected and used by both governmental and private entities. For example, in the United Kingdom, a centralized data protection authority regulates compliance while in Germany, a decentralized network of regulators have been established at both the state and federal level. In the United States, no central monitoring and enforcement entity exists. Regulatory compliance rests primarily on private sector, non-governmental entities, that follow an uncoordinated set of laws governing the use of personal information.

In the United States, there are some sector specific monitoring (e.g. the financial industry); however, no federal privacy monitoring agency exists to enforce regulatory compliance. Monitoring and enforcement are heavily dependent on market-pressure and self-monitoring in those sectors without a formal privacy rules dictating protection of personal information. Monitoring is often accomplished through third party entities such as the Better Business Bureau that serve as watchdogs, provide information about businesses in compliance with privacy laws, and offer information and procedures on how to resolve disputes over the use of personal data (Holtzman, 2006).

### Conclusion

*As long as we live and breathe we'll be paranoid. We always have to be careful, but it isn't going to stop the movement of this technology.*—David Barram

The private sector has used the lack of governmental protection of personal information as an opportunity to develop privacy enhancing technology to sell to the general public. Sometimes individuals do not even know that organizations maintain a record about them. Even more typical is the marketing of privacy enhancing technologies to businesses, who then offer privacy management of personal information as an additional service to customers. However, the business sector has employed privacy enhancing technology more as a means of developing consumer trust and protection of brand recognition and reputation. The greatest incentive for the business sector to use personal data in ethical ways may be the potential for significant consumer backlash. Even though, privacy enhancing technology will always lag behind the technology used to collect and process information. Still, consumers may not want to purchase privacy enhancing technology because of the lack of comprehensive regulations over individual privacy rights and weak legal rights to claims of this protection. In the end, as digital technology such as data mining and processing capabilities increase, so does the emphasis on developing and enacting regulations that provide greater protection over the collection and use of personal information.

## 2020 Prioritizing Information Security for the Next Five Years
CyberWise, University of Fairfax, July 2015

What will be the priorities of information security practitioners in 2020? Organizations are now more complex than ever before and there is no evidence that the next five years will reverse this trend. Companies have adapted to operating in a global and decentralized market economy, placing increasing reliance on vendors, suppliers and contract staff for operations managed previously in-house. They have changed their internal structure to compete better in changing markets and in diverse economic conditions and have learned to leverage new technologies to increase the speed of both communication and business.

This complexity has brought new risks that pose an ongoing security challenge at a time when information security is already arguably at a disadvantage with emerging threats, new attack tactics and new technologies. These developments make information security one of the most dynamic industries around the world. In 2013, there were significant data breaches across multiple industries and governments impacting millions of users. This year brought more of the same. Is this an uncertain future we will have to live in the coming years? Can we accept degraded privacy and security and billions of dollars in lost revenue, damage, reduction in brand value and remediation costs? Organizations need to develop or maintain a robust risk strategy or suffer stunted growth, loss of revenue and legal liabilities.

Future threat environments will force security and risk leaders to create new, adaptive control environments. Given the importance of computer and information security, investment in information security is now recognized as a critical issue by both practitioners and academics alike. Much of the recent information security research has focused on the technical aspects of reducing information security breaches.

Security vulnerabilities, increased dependence on information, and pressure from consumers and regulators requires organizations to spend more on securing information assets. At the same time, increases in information security budgets require proper techniques for evaluating investment decisions pertaining to information technology security. Research devoted to the economic aspects of information security is emerging and has centered on the difficulties associated with the definition and measurement of information technology security costs and benefits.

Better support to business decisions today, information security functions must evolve from a risk-reduction role to a genuine risk management role. The information security function becomes the facilitator of stakeholder risk decisions, providing information and support to the true risk owner to make a decision. The security practitioners require a risk-manager's mindset to help organizations seize emerging technology's opportunities.

The first step in managing a budget is knowing what services the funds will have to support. This may seem simplistic, but it is a step many security leaders cannot complete without a great deal of thought and research. When security programs grow organically over time, it can be hard to keep track of added services without concentrated and continued effort. The same may be true when security leaders must quickly develop their programs based on what is required by regulators or management to provide.

Business managers tend to be reluctant when making

decisions on information security investments as they are associated with uncertainty. Indeed, such uncertainty clouds the managers' vision about information security and many perceive it as a cost or a difficulty, rather than an initiative with strategic values. The uncertainty of information security investments can be reduced with sufficient and quality information, thus allowing managers to justify their strategic decisions and information security improving adoption rate. This approach differs from how information is used to deliver awareness as it adds clarity and encourages managers to balance their decisions on investments, rather than purely promoting the necessity of information security investments.

Security practitioners must assist organization leaders in overcoming this uncertainty by knowing what services the security funds will have to support. Security vulnerabilities, increased dependence on information, and pressure from consumers and regulators require organizations to spend more on securing information assets. Increases in information security budgets require proper techniques for evaluating investment decisions pertaining to information technology security. The security practitioners must build a relationship with organization leaders by proving the information technology security Return on Investment. They need to educate organization leaders about the real security risks to personal identifiable information and intellectual property. Security practitioners have to work side by side with executives to define, dissect, and defend this data and to make a compelling business case for budget. Too many organization leaders spend money on security without taking the time to assess the risk and consequences of an attack on the core business value to the organization. The approach of analyzing security investments, systematic risk assessments, and the economic analysis of the optimal level of security investment based on loss and the likelihood of security violations will make the business management aware of the true business risk.

The proactive security practitioner points out to organization leaders the value of data protection in concrete terms, such as whether or not it generates revenue or helps maintain a competitive advantage. This requires security practitioners to have detailed knowledge of security services, staff and expenditures. Failing to provide an in-depth understanding of where the money goes has implications that extend well beyond the budget. The next generation of security practitioners must know the business and ensure the return of investment from security investment to prepare the organization for the information security future.

Information security professionals must manage their budget responsibly and proactively to protect the organization's high-value and high-risk assets. Such a proactive approach will contribute to the organization leaders' decision for information security investments and to the business integrity and value.

## Genetic Information Nondiscrimination Act

CyberWise, University of Fairfax, August 2015

As a society, we have to ask if we can collect information that reveals individual differences and still continue to treat all individuals the same. How should society treat employers using genetic information in employment decisions, not in order to discriminate, but instead to prevent identity fraud? Apprehension over the potential for misuse of personal health information and genetic data by employers is not entirely new. In response to these concerns, the federal government enacted the Genetic Information Nondiscrimination Act (GINA) of 2008, a law signed by President George W. Bush and designed to provide protection not only from the misuse of genetic data and family health history but also from the initial acquisition of such data.

By genetic information, the Act means test results from the individual or relatives up to and including fourth-degree relationships or the appearance of a genetic disease or disorder in family members. Specifically, the Act prohibits health insurers (but not insurers for life, disability, or long-term care) from requesting or requiring a person (unless he or she is a member of the military) to undergo a genetic test, and from using genetic information to determine enrollment in the insurance plan or the premiums. It also prohibits employers from using genetic information to make decisions about hiring, firing, and other terms of employment.

There is clearly a need for laws like GINA. Genetic testing is now available for more than fifteen hundred conditions, and it generates special ethical concern for several reasons. First, genetic tests expand greatly the power of medical tests. Second, genetic testing threatens our privacy. Testing can produce a tremendous amount of very personal information about an individual from an easily obtained source, usually blood. This information can easily be stored on a microchip or in a computerized database making it all too easy for employers, government agencies, and insurance companies to gain access to a great deal of genetic private information and use it to discriminate against us. Third, the genetic testing of one individual often threatens the privacy of other individuals as well if a person tests positive for an inherited disease or predisposition for the disease. Further, other members within the family, including relatives and children can be at risk of developing the disease or of having a predisposition toward it or of being carriers of the disease-causing gene. An inherited genetic defect is seldom a purely private matter; it is almost always a family matter as well.

Finally, the public grasp of genetic testing is fraught with misunderstanding. Many believe genes are the Holy Grail - the key to the great mysteries of human life, the determining factors in human behavior, and even the explanation of art, morality, religion, and culture. Some say we are nothing but our genes. Moreover, in the minds of many, molecular biology will enable us to predict and control the future.

GINA became fully effective in November 2009 for the purpose of prohibiting the improper use of genetic information in health insurance and employment. GINA applies to employers with 15 or more employees, including private employers, employment agencies, labor unions, and joint labor-management training programs. Under GINA, the Equal Employment Opportunity Commission (EEOC) definition of "genetic information" includes information about an individual's family medical history. The law prohibits an employer from asking about family medical history during an interview, or at any time after the employee is hired. It also prohibits employers from gathering information about the employee through individual genetic tests or genetic tests of a family member. Family medical history is included in the definition of genetic information because it is often used to determine whether someone has an increased risk of getting a disease, disorder or condition in the future. The law has two parts:

• Title I prohibits health insurance providers from discrimination against an individual based on genetic testing. This makes it illegal for health insurance providers to use or require genetic information to make decisions about a person's insurance eligibility or coverage.

• Title II forbids discrimination on the basis of genetic information when it comes to any aspect of employment, including hiring, firing, pay, job assignments, promotions, layoffs, training, fringe benefits, or any other term or condition of employment.

An employer may never use genetic information to make an employment decision because genetic information does not tell the employer anything about someone's current ability to work.

Both GINA and Health Insurance Portability and Accountability Act (HIPAA) protect the confidentiality of genetic information. The privacy rules aim more to prevent discriminatory uses of confidential genetic data by insurers and employers than to protect the confidentiality itself.

Prior to GINA, there was some federal attention to the privacy of genetic information. In 1996, HIPAA was passed to address concerns about discrimination based on a person's health information. HIPAA sets standards for protecting the privacy of individually identifiable health information so as to prevent its inappropriate use and disclosure. HIPAA was the first step toward restricting the use of genetic information by limiting its use in setting insurance premiums and determining a person's eligibility for benefits in group health plans.

GINA requires that the disclosure of protected genetic health-care information be governed by HIPAA. The law also provides participants with injunctive and equitable relief for violations of the confidentiality provisions of GINA. For violations of the privacy provisions of the law, civil monetary penalties of $100 per day up to $250,000 and ten years in prison could be assessed.

GINA records must be retained by the employer for a period of one year from the date of the making of the record or the personnel action involved, whichever occurs later. In the case of involuntary termination of an employee, the personnel records of the individual terminated must be kept for a period of one year from the date of termination. Where a charge of discrimination has been filed, or an action has been brought by the EEOC or the U.S. attorney general against an employer under GINA, the employer must preserve all personnel records relevant to the charge or action until final disposition of the charge or the action.

The passage of GINA has important implications for patients and their families. The feasibility of a National Health Information Infrastructure (NHII) becomes increasingly greater as communication and information sharing technology becomes increasingly sophisticated. The three most critical points to address in creating such a system are:

- confidentiality of patient records/security breaches
- viability of large-scale data interchange and
- the ability to synchronize national standards and regulations.

The increasing use of electronic health records over the past decade or so has brought up numerous ethical concerns about the confidentiality of patient data. Many Americans are afraid that their personal and "private" medical information might be used against them to limit their insurance coverage or ruin their chances of a getting a job or promotion.

As organizations have begun to rely more and more on electronic health records, the ability to keep people with bad intentions at bay has decreased. It is a lot easier to guard a file cabinet from an individual trying to break in than it is to guard an entire electronic database from potentially hundreds of people who want to hack in for whatever reason (reporters, insurance agents, etc.). But it is a lot cheaper to store health records electronically, and it also takes up a lot less room, so these benefits often outweigh the security disadvantages.

Human genome research is creating new opportunities for a more individualized approach to the screening, diagnosis, and treatment of rare and common diseases. Until recently, many people were concerned that information resulting from their genetic testing would be used against them by their insurers and/or employers. With the passage of GINA in May 2008, U.S. citizens now have federal protections against insurance and employment discrimination based on their genetic information.

Although GINA has been the law throughout the land for over five years, the public is largely unaware of its provisions. It is the duty of public officials and health care professionals, to inform American citizens about their choices, options, and right of genetic testing. With the advancements in technological data gathering, it is equally important for Information Security and Health Care Information Security and Privacy Practitioners to understand the benefits and necessity of genetic testing, as well as the rights and protections secured by GINA.

## 2001 Unique Challenges

I was in the Joint Staff Information Technology Operations Center, Pentagon when the airplane hit September 11, 2001. This essay was my way to overcome the chaos of the day. Today this nation is debating this important question of civil liberties and to reconfigure some aspects of democracy for individual safety.

December 1, 2001

There is a unique challenge in defending the homeland. For the first time in our history, the Armed Forces may not protect us from a foreign threat. Today, the "first to fight" may well be a police officer, a volunteer firefighter, a hazardous material technician, a nurse or even an information security technician. The United States must view homeland defense as a partnership among federal, state, local, and private-sector organizations. Homeland Defense must reorganize vertically – on the federal, state, and local levels - and horizontally in coordination with the executive branch.

The nature of terrorism is changing. While low-cost kidnappings and bombings have had the order of the day for decades, high-tech attacks on large numbers of people on a nation's infrastructure are increasingly the preferred tactics. This paper discusses some of the issues we need to address in the search for long-term solutions.

## Reexamine Civil Liberties

In peacetime, Americans enjoy all the protections of citizens' right resulting from decades of constitutional interpretation and amendment. In wartime, national security interests can override such rights. Now, the advance of terrorism is pushing us into an ambiguous situation between these two poles, raising philosophical and practical questions.

Americans must balance idealism with reality and recognize that terrorism has become a national security issue. The freedoms of democracy, as we have known them, may need modification to permit stronger action for thwarting terrorists. The U.S. system of justice has been sufficiently resilient to adjust too many other national security dangers of the past.

Civil liberty advocates are rightly concerned over evidence that at times, even in the United States, government employees are corrupt. Large segments of the American public see abuse of government power as a greater enemy than terrorism. Thus, as a top priority, law enforcement agencies must put their houses in order before trying to expand their reach. From FBI agents to state troopers, from customs inspector to municipal police, all public employees involved in law enforcement must be held to a higher standard of conduct than employees in any other sector. The U.S. Armed Forces have their own code of conduct, the Uniform Code of Military Justice, which supplements the legal framework to which the average citizen abides. Law enforcement agencies should have an even more stringent professional code for their employees. Many aspects of such a code, such as random drug testing, financial disclosures, and confidentiality guidelines, are already in effect at some federal agencies. These structures, however, should be expanded and given sharper teeth through swifter and more severe penalties for violations of the public trust.

With a no-nonsense code in place that reduces the temptation of corruption, law enforcement officials would be able to argue more effectively for broader powers to combat terrorism. And, most importantly, many doubting Americans could eventually be swayed to have greater confidence in their government.

### Need for Public Cooperation

The public and private sectors need stronger alliances and interagency coordination to stave effectively off terrorism, particularly super terrorism. Law already mandates some types of cooperation and have established precedents. For example, companies file reports with the Department of Commerce concerning exports of sensitive items. Financial institutions keep records of all transactions exceeding $10,000. Chemical companies report activities that threaten health safety.

Appealing to the Americans' sense of responsibility to provide information simply because it is the right thing to do has not been fully explored. Just as banks should have a know-your-customer policy, companies, and other organizations handling dangerous materials should know their customers. As a case in point:   Once the U.S. government approached it after the Oklahoma City fertilizer-bomb incident, the Fertilizer Institute encouraged its members to report suspicious purchases. Other suggested approaches include:

The retailers and distributors of chemicals that can be used as explosives should be embraced as full partners by enforcement agencies to help keep these chemicals out of dangerous hands.

The owners and operators of power, water, transportation, and communication systems should be in continuous contact with federal and local authorities concerning the security of these vital systems.

American importers of goods from drug-producing or drug-transit countries should know the histories of their shipments. They should know the reliability of loaders and transporters along the entire shipment route, as the narcotraffickers – often financiers of terrorists – attempt to hide drugs in shipments of textiles, furniture, toiletries, and all manner of goods.

Multinational corporations have a global presence and share responsibility and cooperate with governments and law-enforcement agencies.

Multinational corporations need to operate within the laws of every country they enter and treat workers better than the local law requires. Ideally, they should pay workers on the same scale wherever the same work is performed.

In own neighborhoods, Americans should be alert to strange developments around them and be less hesitant to inform authorities of suspicious activities. The threat of terrorism should be taken seriously at every level. Americans should lobby Congress and federal agencies to implement strategies against terrorism, such as protecting public buildings, maintaining stockpiles of vaccine against anthrax and other likely biological terror agents, and training doctors how to recognize and treat victims of chemical and biological weapons.

### National Campaign on Public Preparedness.

Preparedness begins at home (that is, preparedness is a responsibility of all citizens). All citizens should try to protect and provide for themselves and their families in the event of a natural or manmade disaster. These responsibilities include personal preparations (such as food, water, and emergency supply storage) as well as support to reasonable preparedness programs.

An uninformed (or misinformed) public is one of the greatest dangers our society faces. Provide reliable disaster related information and to promote public awareness of civil defense issues. The preparedness program should promote dual-use preparations wherever possible so that the cost of preparing for disasters is offset by normal day-to-day utility (such as with the Swiss concept of using parking garages for blast shelters). Also, the dual-use concept can extend into the dimension of ensuring that preparations for natural disasters are done in such a way that preparations for war or terrorism are also addressed.

The government should not hold a population hostage to known dangers, which occurred with policies like Mutual Assured Destruction (MAD) and is at best irresponsible and goes against the Constitutional mandate to provide for the common defense. The best defense against manmade disasters, such as terrorism and war is a credible deterrence combined along with the strength to survive.

### A new role for NATO

NATO could certainly craft a new article placing counter-terrorism on its security agenda. Closer integration of intelligence functions is a logical first step, followed by joint training, and response exercises. In time, counterterrorism could become a major NATO undertaking as the European countries recognize that the NATO military infrastructure can provide important support for law enforcement agencies.

### Addressing Root Causes

Historically, much of terrorism's fuel has been pumped from the wells of ethnic and religious conflict. This will be the most difficult facet to fight. However, modern mutations of terrorism are also inextricably linked to overpopulation, food shortages, and environmental devastation. Much violence can be traced to the struggle between haves and have-nots, with the boundaries between war and terrorism increasingly blurred.

We should consider a Cooperative Terrorism Reduction

Initiative (CTRI). The key is the word *cooperative:* not foreign aid but a program that benefits both sides—giver and receiver—and depends on contributions from both sides. This program would operate in tandem with the many quick fixes we currently support to thwart terrorism in the very near term. The CTRI program would be directed at creating major changes in those countries of greatest concern—perhaps beginning on a pilot basis with countries where the task seems easier, such as the Philippines and Peru. In time, the program could target difficult countries with more complex problems, such as Colombia, Algeria, and Pakistan. The funding levels must be substantial—billions of dollars annually to start—since the task is to promote legitimate economic development as a real alternative to crime and terrorism.

With much more financial support at stake, the participating countries would have stronger incentives to attack government corruption, eradicate terrorist activities based within the country, prohibit imports of sophisticated weaponry, eliminate drug trafficking, and stiffly penalize financial institutions that launder money. At the first sign of corrupt use of the funds, the programs should cease without debate.

A CTRI program should operate with several cardinal principles. Job creation would be the cornerstone of the program. Those young people with the guile to thrive in the barrios and ghettos are precisely the type of youngsters who, given the chance, could be successful in industry and commerce rather than crime. In addition, local organizations committed to improving the standard of living would have the means to provide constructive outlets for both leadership skills and aggressiveness.

Complete openness, or "transparency," in all actions involving advanced technologies would also be essential. For example, development of medical, veterinary, and pesticides capabilities should be strongly supported in exchange for transparency of such activities if they could be used for bioterrorism or chemo terrorism. Finally, imports of armaments—beyond appropriate self-defense needs—should be minimized or banned as a precondition. Of course, the devil is in the details. Difficulties in implementing effective programs have almost always been the Achilles' heel of the Agency for International Development. Indeed, AID should not implement this program; the Department of State would better sustain the cooperative character of the program. Also, a new breed of American program managers is essential. No longer should American experts unilaterally design projects in other countries. Team players should work in partnership with recipient organizations. In the long run, the investment in a CTRI program would be small compared with the costs of defending American interests on an ad hoc basis from constant eruptions of terrorism.

### Proper Judgment

Some 1.2 billion people worldwide struggle to survive on $1 day or less, 1.2 billion people lack access to safe drinking water, and 2.9 billion have inadequate access to sanitation. About 150 million children are malnourished, and more than 10 million children under five will die in 2001 alone. At least 150 million people are unemployed, and 900 million are "underemployed"-- contending with inadequate incomes despite long hours of backbreaking work. Poverty and deprivation do not automatically translate into hatred.

The United States and the other industrial nations should build an international coalition to provide everyone on earth with a decent standard of living. A 1998 report by the United Nations Development Programme estimated the annual cost to achieve universal access to a number of basic social services in all developing countries: $9 billion would provide water and sanitation for all; $12 billion would cover reproductive health for all women; $13 billion would give every person on Earth basic health and nutrition; and $6 billion would provide basic education for all. These social services expenditures pale in comparison with what is being spent on the military by all nations--some $780 billion each year.

By choosing to mobilize adequate resources to address human suffering around the world, we must all understand that in the end, weapons alone cannot buy us a lasting peace in a world of extreme inequality, injustice, and deprivation for billions of our fellow human beings.

## Oil

The increased oil consumption within the country has made America heavily dependent on oil imports from the Arabic world. Such dependence makes the country vulnerable to terrorist threats aimed at the production and transportation systems supporting "the American way of life." Thus, the concept of a threat to the critical infrastructure needs to be expanded to include the possible attack of external resource networks established to supply the internal infrastructure with internally unavailable or scarce resources. The threat to America's oil supply can arise from two general scenarios:

1. Physical damage to pipelines through terrorist attacks on the territory of US-friendly Regimes. Therefore, it will make sense for the United States to remain in Afghanistan until political guarantees are in place to ensure the safety of the oil pipelines going through the country.

2. Disruption of oil transmission or destruction of wells by regimes hostile to American

3. Interests. The threat in this scenario may come from a Taliban-like regime taking over the control of Saudi Arabian oilfields, for example. Many of the Taliban leaders, including Bin Laden, have been dislocated and are

likely to focus their attention to Saudi Arabia, whose government, from Bin Laden's point of view, is a puppet of American interests. Taliban-like regimes do not necessarily operate on market principles or in a self-interested fashion. A religious vow to punish the West for its consumerism may lead to irrational decisions like the destruction of wells.

## The Media

The media and its enormous influence on "the American way of life" can be easily underestimated. The first thing many Americans did on September 11 was to turn on their TVs for information and guidance. Imagine the effect if there had been nothing on the air, or even worse if the air had been taken over by the attackers. The control of the media and their transmission channels pose another potential threat. Like oil, strategic parts of the system can be damaged outside of the territory of the United States, which raises additional challenges in the defense of critical infrastructure.

## Education and Reliance on Foreign Intellectual Capital

The American educational system is consistently failing to train enough American-born scientists and researchers. As a result, private companies and state-run research centers working on military contracts continue to hire foreign-born specialists. While almost all of these foreign scientists eventually become American citizens, it is hard to determine where their allegiances lie. The argument is not that most of them are likely to betray their new-found home-country; many of them, however, still have family in their native lands and are potential blackmail targets. Some take advantage of the privileges and benefits given by the citizenship but never truly become "Americans." Others, because their allegiance is not to the United States, are likely to move to a different employer or a country offering better more lucrative employment conditions.

### New Weapons

The traditional weapons of the terrorist were cheap, readily available, and one that would do the job. They ranged from the knife to the small automatic rifle to the plastic explosive. But now that they have the money and technical facilities of whole countries to back them, terrorists have a whole new range of weapons to use against a completely new set of targets:

**Stinger hand-held rockets.** Small, light, and powerful, Stingers can be used, with almost no training, to knock out an airplane.

**Computer viruses.** Viruses are programs that destroy data stored in a computer's memory. It is relatively easy for a programmer to write a program that will tell the computer to erase its memory or otherwise render its programs useless. What makes these viruses so hard to trace is that they often contain timing instructions, so that it might be months between the time a virus is introduced and the time it is triggered. The programmer simply embeds the codes into an otherwise harmless program and then tries to get the program into a victim's computer.

**Electromagnetic pulse generators.** Put a pulse generator on the power line to an important computer, and the pulse will wipe out the data in the computer's memory.

**Chemical and biological weapons.** Chemical weapons range from old-fashioned poison in the water and nerve gas to a new Liquid Metal Embitterment agent (LME). Applied with a felt-tip type pen, LME is a clear, invisible substance that changes the chemical structure of a metal so that it is no longer resilient and flexible. The result: The metal can fracture under stress. Trucks, airplanes, or bridges would be vulnerable to catastrophic failure without advance warning.

The standard chemical weapons are frightening because they can be concocted from cheap and readily available materials. Common pesticide and fertilizer components can be used to make poison gas while deadly chlorine gas can be obtained from the electrolysis of ocean water. Chemical and biological weapons may become the "nuclear weapons" of small countries since they are weapons of mass destruction that are easier to come by than nuclear weapons. Perhaps industrialized nations should consider banning from their schools and universities students from developing countries manufacturing biological or chemical weapons since one cannot produce these without a substantial technical education.

### The Coming Rise in Domestic Terror

The biggest change in low-intensity conflict in the next decade will be an explosion in the incidence of domestic terror in the United States. Security measures currently in place around the world will help slow the spread of international terrorism, but the frustrated in the United States will begin to take their cue from terrorists abroad. Some of the groups that have already begun terrorist-type actions include:

**Antiabortionists.** Planned Parenthood offices and women's health clinics that offer abortion have already been bombed. Attacks will get worse because religious groups believe that God's law puts them above civil law and other people's rights.

**Drug dealers.** Terrorism by drug dealers will aim at breaking the resistance of city and state governments and law enforcement agencies. It will be sponsored by organized crime and "disorganized crime" —the kind of crazed violence observed in crack users that stuns those who haven't seen it. The international terrorists may inspire the methods, but the organized-crime armies have been trained and weaned in Colombian drug wars. The U.S. government can muster tremendous resources to fight terrorism within its borders, and national leaders have always had to deal with threats from other countries. Few city mayors or state governments, however, are psychologically prepared or fiscally and strategically able to deal with sustained pressure from terrorism.

**Counterterror from the right.** In Colombia, people who got tired of police inability to deal with terrorism have formed death squads to "help out" the police. The same thing could happen in the United States.

**New Targets.** Many vital industries and resources are staggeringly vulnerable to attack. Even if there were the will to do so, it would be expensive and inconvenient to guard every office and factory. But some changes will have to be made to reduce their vulnerability to crippling terrorist attacks.

Computer attacks already account for some 60% of all terrorist attacks in the world. Twenty-four computer centers were bombed in West Germany in one year. Italy's Red Brigades and France's Action Directed have both targeted computer systems in Europe. It is only a matter of time before someone takes advantage of U.S. computer vulnerability.

France and Spain have already felt the terror when antinuclear groups broke into nuclear plants to protest their operation. These groups only shut down the power, but there is the threat of reactor meltdown and widespread radiation damage due to terrorist attacks on these facilities.

Only two pipelines supply virtually all the natural gas to the northeastern United States. Both are regulated by high-pressure pumps that are manufactured in foreign countries, with replacement times of more than a year. Both pipelines are essentially unprotected. Two bridges, one over the Ohio River near Cincinnati, the other over the Potomac River near Washington, D.C., handle all the north-south railroad traffic in the eastern United States. Neither is guarded.

Fewer than ten regional switching stations control virtually all the telephone communications in all the large cities of the United States. In May 1988, one of these stations caught fire in Hinsdale, Illinois, a suburb southwest of Chicago. It is an automated facility, with a single watchman on site and an overseer who monitors warning lights in a facility 100 miles away. By the time the technicians believed their warning lights and summoned help, the station was destroyed, plunging one-third of all Western Chicago phones into silence. Because the station was between O'Hare Airport in Chicago and the regional air traffic control center in Aurora, Illinois, O'Hare was without much of its air traffic controls for hours. Even with round-the-clock work crews, it took three months to repair the station. Phone service was not fully restored until August. With no phone transmission of any kind, many computer networks also went down. Only cellular phones using satellite transmission, such as car phones, were unaffected.

The telephone switching station fire points to a vexing problem for the increasingly high-tech industrial United States. There are fewer and fewer individuals who monitor the safety of the nation's highly automated industrial facilities. The human overseers who are there have enough experience with faulty equipment that they often distrust malfunction signals. The telephone overseer in Illinois disregarded his first "fire" signal because there was an electrical storm between his office and the switching station, and he knew that could cause a false display. Another problem is that these automated systems often have security systems that are tied to regular electric and phone transmission lines. If something happens to the electrical or telephone system, the security people are often "left in the dark."

**Know your enemy.**

It is the oldest military axiom in the world. The first step in defeating terrorism is figuring out how terrorist organization operates. Think of terrorist of dispersed network nodes – linked, as PCs are, to one another and databases of information. The network uses the Internet for real-time dissemination of instructions to wage its particular brand of asymmetrical warfare – favoring an attack on a soft civilian target like an airliner over a direct assault on US forces in the field. We already know a few things about such networks; structurally, al Qaeda is similar to Colombian and Mexican drug cartels, which also feature small, nimble, and dispersed units capable of penetrating, disrupting, eluding, and evading.

The US military is woefully unprepared to fight a war against such an enemy. Transnational terrorists have shown it's possible to swarm together swiftly, on cue, then pulse to the attack simultaneously. Simply dropping bombs on Afghanistan will do little against this kind of a decentralized foe. To win, this network must be isolated, node by node. The US intelligence systems should be redesigned to anticipate how this new enemy thinks. The US must build its network – a quicker, more diverse, populous, and powerful organization that includes military and nonmilitary organization around the world – to wage a full-on net war.

**Infrastructure strategy.**

The concentrations of assets and people make valuable targets. Lesson one: disperse vulnerabilities – which means breaking up everything from the energy industry to air travel and operating systems. Distribution and diversity throughout the national infrastructure changes the risks and decreases the interdependencies, making the infrastructure more robust and reliable.

If a city's water supply relies on a single river, or if all of its electricity comes from a grid that can be brought down at a single switching station, then there is a concentration of consequence in the river or the grid, which makes them a high-value targets. Building diversity and distribution into your system changes the infrastructure and keeping things spread out reduces vulnerabilities.

The eight infrastructures are telecommunications, electrical power systems, gas and oil transportation and storage, water supply systems, banking and finance, transportation, emergency services, and continuity of government operations. Interdependencies have been created among these eight critical infrastructures, making them more vulnerable. Without electric power, other critical infrastructures, such as telecommunications and banking and finance, cannot function. Interdependencies exist among the infrastructures so that minor but coordinated attacks in several key areas could result in a phenomenon of cascading failures.

If arguments over national defense could reshape America's infrastructure in the 1950s, why not again? The 43,000 miles of road network officially called the System of Interstate and Defense Highways represent a defining piece of American infrastructure. Eisenhower wanted to generate a lot of employment after the Korean War, part for economic reasons and a way to move armies across the country quickly to get people out of cities threatened by nuclear attack.

Decentralization of energy sources and building power generators of the needed size as close as possible to where the power is needed. For example, a series of micro turbines based on the same technology as jet engines that can run on a variety of fuels, including natural gas, methane, or gasoline could be d nearby a new suburb or new town. This system could eliminate the need to string electrical cable across the countryside and could save a lot of the seepage waste in distribution, not to mention eliminating the many miles of power line that could be potential targets.

Also, the widespread adoption of hydrogen fuels cells as part of a decentralized energy infrastructure would be a further means to the same end. Hydrogen is a way of storing energy and thus increasing the period during which supply and demand may be matched.

Biological attacks are different, viruses and bacteria, gain new targets as it goes on and can be diffuse in its essence, but its impact building up over time. There are ways to lessen its potential impact; you must become aware of what is going on as quickly as possible. A fast response is another defense that a distributing infrastructure can help provide. To detect an attack quickly requires a rapid well-distributed public health information system. An early warning system to detect and analyses reports of symptoms from emergency rooms revealing unusual patterns of disease. Also beside speed, such systems would allow better coverage of outbreaks in other countries, a basic requirement for any bio warfare early warning system.

Stockpiles of vaccines and antibiotics need to be built up and made available at a range of locations in a range of countries. Another distributed response to biological attack in partial immunity. There are already vaccines against most plausible bioweapon agents. If a small percentage of health workers were to be choose to be vaccinated against one or some of these diseases, then a reservoir of manpower would always be on hand in an emergency, ready to help with the vaccination of others or to do whatever else was necessary in places where infection was rife. You cannot vaccinate everyone against everything; but if some people are vaccinated against most things, and you know where to find them, their distributed immunity could be a powerful asset.

### Reorganize

There is a need an integrated federal effort, routine threat assessments, and reconfigured national security architecture. The variety of federal agencies and programs devoted to counterterrorism remain fragmented and uncoordinated, with overlapping responsibilities but no clear focus. Regular assessments of the terrorist threat, both foreign and domestic, as it exists today and is likely to evolve in the future. Restructure the U.S. intelligence community to counter the terrorist threats of today and tomorrow rather than yesterday. Our national security architecture is a cold War-era artifact, created more than half a century ago to counter a specific threat from a specific country and a specific ideology.

The inability to effectively reform institutions and systems will cause decline. The greater challenges are to reform our institutions and to go beyond simple bureaucratic fixes and to restructure radically our foreign and domestic counterterrorism capabilities. Just as the narcotics problem is regarded as so great a threat to our national security that we have a separate agency — the U.S. Drug Enforcement Agency —specifically dedicated to counternarcotics, the Office of Homeland Security should be the structure in which counterterrorism efforts all domestic and defense agencies can be coordinated.

### Conclusion

It is impossible to eliminate all the motives that drive terrorism. Confronting the underlying social and economic motivations could reduce the frequency and magnitude of future incidents. No single government agency or program alone will reduce terrorism, but combining efforts in aggregated initiatives can be very effective. We do not have adequate threat estimates, technical assessments, or net assessments to guide our leaders. We must overcome the complexity involved in these assessments and have wide interagency cooperation on the federal level and also include state, local, and private sector organizations.

Finally, civil liberties are important for a democratic society; the time has arrived, however, to reconfigure some aspects of democracy. Minor fine-tuning now can reduce the need for major changes in the future. It is simple to increase the pay of airport security personnel, but they must be willing and trained to search for weapons because of pride in their job and, more importantly because they are proud to be Americans or living in America.

# National Patient Identifier and Patient Privacy in the Digital Era

*by Tim Godlove, PhD, and Adrian W. Ball, PMP, ITIL, CPHIMS*

Information Security Management Handbook, Sixth
Edition, Volume 7, August 2013

Information cannot be shared among healthcare providers if systems are not interoperable. The current lack of standardized technology not only prevents integration and interoperability of information systems but also severely limits the ability of Electronic Health Records, often referred to as EHRs, to safely facilitate continuous, informed care across healthcare settings. In other words, healthcare information systems need to talk to one another.

Assigning everyone a universal patient identifier, or UPI, would improve a doctor's ability to share information and make it easier for hospitals to differentiate one William Smith from another. However, a universal health ID system would empower government and corporations to exploit the single biggest flaw in healthcare technology today: Patients would not have access to who sees, uses, or sells their data.

## Overview Electronic Health Records

Until the latter half of the 20th century, physician and practitioner notes, medical tests, and medical images were written by hand, maintained in a tangible medium, and stored in a physical location. Though there were many technical problems and other issues such as lack of system interface and non-standard vocabularies, health information stored in an electronic format has been around since the 1960s, predominantly at academic and research medical centers. Beginning in the 1960s, patient information has slowly progressed toward being wholly recorded and stored in EHRs. Both President George W. Bush and President Barack Obama pushed for digitizing U.S. health records by 2014. To encourage medical institutions across the country to convert to an EHR system, Congress earmarked $19.2 billion in incentives in the American Reinvestment and Recovery Act (ARRA).

According to the Health Information Management Systems Society (HIMSS) 2011, EHRs are longitudinal electronic records that contain patient health information generated from multiple clinical care delivery settings. Most EHRs contain two types of data—data scanned into a graphic file and electronically created records that fall into four broad categories: prescription orders, orders for tests, tests results, and physician notes. Information in EHRs typically includes patient demographic information, clinician treatment notes, presented problems, lists of medications, medical history notes, laboratory and radiology reports, and other healthcare related information. The EHR is intended to streamline patient-care workflow and provide access a complete record of health care by way of interconnection with one or more EHR systems, decision support, quality management, and outcome reporting.

Those who support the move toward a digital medical information access and storage system argue that such a move would result in significant savings to a health system with upwardly spiraling costs. The increased accessibility for clinicians to important health information about the patients they treat, will deliver other savings and benefits, including reduction in needless duplication of medical tests and procedures. Furthermore, access for clinicians to a more complete history of the patient's care over time, which will help decrease medical errors. In a nutshell, EHRs are designed to improve the overall efficiency and quality of patient care at less cost and with fewer errors resulting in injury or death.

Doctors and other healthcare providers are not the only individuals, groups, or even institutions that maintain EHRs. Insurance companies and third-party payers such as Medicare and Medicaid maintain electronic patient health information. While EHRs must comply with regulations mandated in the Health Insurance Portability and Accountability Act (HIPAA) limiting access to healthcare clinicians, other institutions in possession of personal health records are not required to comply with HIPAA, even though those institutions often control the information contained in the personal health record.

The conversion to total use of EHRs is not without its critics. The primary objection against the use of EHRs is a concern that the security measures in place will not sufficiently protect the privacy of patient information or deal with breaches of security, patient medical identity theft, and unlawful access. Another concern is the use of medical information by employers, insurance companies, and any other individual or entity wishing to use the information for their benefit and to the detriment of a patient. In addition, concerns exist about the indirect use of medical information without a patient's consent such as for research, provider certification, public reporting, marketing, and other commercial activities. Clinicians have expressed concern over complete digitization of medical records due to the risk of not having access to information when needed because of problems with the technology and working with technology experts who do not fully understand the need to protect patient privacy.

### National Patient Identifier and Universal Patient Identifier

Within the healthcare delivery system (examples include physicians, insurance companies, managed care organizations, hospitals, or pharmacies); patients often have several different identifiers. During their lives, patients move to different locations and access a variety of healthcare providers who maintain records of the health care provided. Use of a unique identifier is intended to simplify access to electronic health records and thus improve the quality of care, reduce administrative costs, and decrease injury and death caused by medical errors. The purpose of a UPI that each patient would be assigned is to eliminate the use of different identifiers for the same patient across different health care providers. The UPI would allow for better continuity of care, accurate record keeping, improved follow-up and preventive care, correct and prompt billing and payment for services, decreased waste, and detection of fraud and misuse of patient information.

Unlike a social security number that has been used an as a unique individual identifier in for a broad range of purposes such as school records, employment, Internal Revenue Service identification, or accessing financial accounts, a UPI would be a unique number for accessing only medical records (HHS,1998). HIPAA legislation supports the creation of a unique identification system, but such a system has not been implemented because of concerns of patient privacy and security of medical information. In the last ten years, adoption of EHRs has expanded considerably. According to Hillestad, et al. (2008), who conducted a study for the RAND Corporation on the use of electronic health records and UPI, the cost to create a national identification system would be about $11 billion.

The Healthcare Information Management and Systems Society put considerable effort into promoting the use of a UPI. In 2011, HIMSS argued that the use of a national identifier would increase the ability to both access and secure health information. In addition, HIMSS made three broad recommendations to Congress in 2011 in its support of a nationwide patient identifier (HIMSS, 2011). First, HIMSS recommends in its report that Congress continue supporting acceptance and implementation of health information technology, arguing that a National Patient Identifier used in association with EHRs will promote better care, reduce errors, and increase billing and payment efficiency. Second, HIMSS recommends that Congress continue supporting investment in the Medicare and Medicaid EHR Meaningful Use Inventive Programs that will continue the efforts made by both Presidents Bush and Obama to move to a national EHR system. The third recommendation by HIMSS is that Congress remove any barriers to creating a national health identifier included in the 1999 Omnibus Appropriates Acts that prohibits the federal government from using funds toward creation of a system that uses a national health identifier (HIMSS, 2011). Finally, HIMSS points out in its report that implementation of a national health identifier is not synonymous with issuance of a national identity card but is a means of linking a unique electronic health record with the patient.

### Privacy and Security concern over using a National ID or Universal Patient Identifier

While use of digitized health records offers the benefit of providing clinicians with easy access to and sharing of medical information with other clinicians, ease of storage, and speed of transmission, EHRs make this information vulnerable to security breaches. Other vulnerabilities are exposed when businesses use private medical information for their own purposes and profit without notification or permission obtained from patients. Although, the implementation of EHR has many benefits, one area of considerable concern is the protection of patient privacy and dealing with security breaches and patient medical record theft. The Clinton Administration instituted federal regulations to safeguard EHR privacy. The Bush and the Obama administrations continue the effort to implement a total shift to EHRs.

Health information stored in EHRs is a valuable commodity to criminals who sell such information on the black market or use it to commit Medicare fraud. Between 2010 and 2011, there was a 97 percent increase in health data breaches in the United States (Manos, 2012). Since 2009, 19 million people have been affected by health information breaches that occurred across all 50 states (Manos, 2012). Part of the problem is the lack of security that prevents health information from being stored on unencrypted laptops and other portable storage devices (Manos, 2012). In addition, there is little oversight in protecting patient privacy among health care organizations that disclose patient information to their third-party business associates (Manos, 2012).

## Arguments supporting the use of Universal Patient Identifiers

Those who support implementation of a UPI system argue that such a system would increase the level of security. According to Hillestad, et al, (2008), use of system check codes tied to the UPI would guard against the number of input errors as well as lessen the chance of accessing and entering data into the wrong patient electronic health record. Further, use of a UPI would improve the ability to store and retrieve records across different systems and would enhance the quality of care by reducing repetitive and unnecessary procedures.

Proponents of a UPI system point out that without a unique identifier, most health systems and healthcare providers use a technique called "statistical matching," which uses attributes such as name, birth date, sex, and social security number to retrieve a patient's medical record. However, statistical matching can result in the retrieval of incomplete EHRs about eight percent of the time. In addition, statistical matching increases the risk of privacy breaches because a great amount of personal information is used during the records search process. Hillestad, et al. (2008) argues that the use of UPI makes implementation and maintenance of security protocols easier than using the statistical matching system used today.

The President's Council of Advisors on Science and Technology (PCAST) Report to the President Realizing the Full Potential of Health Information Technology to Improve Healthcare for American: The Path Forward of December 2010 does not support a requirement for UPIs or the creation of Federal databases of patients' health information (Holdren and Lander, 2010). In addition, the report does not explain a proposed approach to eliminate the need for a UPI. Another major hurdle with EHRs is the incorrect record linkage, both for patient care and research needs. How will the systems know the appropriate linkage between medical records for the same individual, particularly across a myriad of healthcare organizations, providers, and EHRs that may contain information on the same person? Relying on identity resolution technologies and probabilistic person matching algorithms are imperfect, and do not resolve to identify individuals with sufficient certainty to be neither used in healthcare nor biomedical research (Niland, 2011).

A Health Information Exchange (HIE) can be thought of as a database of databases where the databases are the data sources (e.g., EHRs and Enterprise Master Patient Indexes (EMPIs)) exposed by the participating Health Information Organizations (HIOs). The database is the collection of those exposed data sources. On the macro level, HIEs support three sequentially executed database queries: patient-match, available-patient-documents, and patient-documents-retrieve. An HIO first executes the patient-match query to determine if other HIO databases (e.g., EMPIs) contain a matching patient. The patient-match query only returns those matching patients who have agreed to participate in the HIE. An HIO then executes an available-patient-documents query to determine which documents (e.g., a Continuity of Care Document (CCD)) the matching patient(s) has consented to share with the querying HIO. A patient may have some medical information (e.g., a mental health issue or sexually transmitted disease) they do not want shared with some or all of the HIO's participating in the exchange. Finally, an HIO executes a patient-documents-retrieve query to retrieve all or some of the documents the matching patient(s) has consented to share with the querying HIO. E.F. Codd, the inventor of the relational database model which is used to manage the majority of today's electronic data, identified twelve rules for defining a relational database. Rule number two, Guaranteed Access, stated that all data must be accessible without ambiguity (Codd, 1985). This rule is essentially a restatement of the fundamental requirement for primary keys. Today's statistical matching algorithms violate this rule and can return false-negative and/or false-positive patients in response to a patient-match query. This violation creates data-management issues with negative patient safety, privacy, security and public trust consequences. HIEs could address these issues by adopting a UPI to support unambiguously identifying patients across the HIE database allowing patient-match queries to return either a unique matching patient or a "no patient found". Assignment of the UPI could be voluntary to address most of the objections that have prevented the adoption of a patient identifier but should be mandatory for patients wanting to participate in an HIE.

### Arguments against issuing Universal Patient Identifiers

Senator Tom Coburn of Oklahoma, who is a practicing

physician, raised concerns that security measures at this time are not sufficient to protect information stored in EHRs from cyber-intruders or others who would unlawfully obtain and use personal health information (Sterstein, 2011). President Obama in the American Reinvestment and Recovery Act allocated more than $20 billion dollars for use in converting existing paper-based medical records to a digital format and to implement EHRs. Senator Coburn points out Chinese cyber-attackers have breached security measures and obtained sensitive information—technology that could be used to hack into and obtain private medical information worldwide. Legislation generally does not include provisions for reporting data breaches involving health information. According to Sterstein (2011), "All but one of the several pending Senate bills that would mandate data-breach notifications excludes health information." Senate Bill 1535 (S-1535) that Richard Blumenthal of Connecticut sponsored does provide support for disclosing breaches involving health performance (Sterstein, 2011). In 2011, 73 individuals were indicted for healthcare fraud and medical identity theft for stealing the identities of doctors and Medicare beneficiaries and using this information to submit $163 million in false billing. In June 2011, an individual who was a defendant in an administrative hearing was able to access sensitive medical records of Arizona state government employees involved in litigation with the defendant (Sterstein, 2011). Those are just a few examples that highlight weaknesses in current security systems to protect patient identity and health care information.

### Universal Patient Identifier Implementation

The UPI System could be included within the Identity Ecosystem (IE) described by the National Strategy for Trusted Identities in Cyberspace (NSTIC) (Baker, 2011). The NSTIC IE is made up of the following:

1.    Trustmarked organizations that have met the requirements of the IE as   determined by an accreditation authority,

2.    individuals engaged in an electronic transaction,

3.    subjects of the transaction, and

4.    identity providers responsible for establishing, maintaining and securing the

5.    subject's identity.

In an HIE, the trust marked organizations are the HIOs that have complied with the overarching set of interoperability standards, risk models, privacy and liability policies; requirements, and accountability mechanisms of the HIE (Baker, 2011). The individuals are the participating HIO's HIE users (e.g., physicians) and the subjects are the patients who have consented to participate in the HIE. The Voluntary UPI (VUPI) is the digital identity assigned by the HIE's identity provider. An identity provider would be responsible for establishing, maintaining, and securing the VUPI for those patients wanting to participate in the HIE. The VUPI could enhance patient control over the privacy of their information, improve the quality of medical care and the efficiency of its delivery, reduce medical errors related to misidentification of patients; decrease incidents of healthcare-related identity theft and help control healthcare costs as a result of these impacts (Hieb, 2012).

The American Society for Testing and Materials (ASTM) standards E1714 (Standard Guide for Properties of a Universal Healthcare Identifier) and E2553 (Standard Guide for Implementation of a Voluntary Universal Healthcare Identification System) could be leveraged to implement a VUPI System. According to the ASTM standards, the VUPI System would not include a central database of patient demographic information. Instead, it would integrate with the HIOs' EMPI which would in turn assign a VUPI (Dimitropoulos, 2009). A production VUPI System based on these ASTM standards became operational in May 2011 as part of a pilot with the Western Health Information Network (Hieb, 2012).

The following example describes the assignment of a VUPI in the context of a for-treatment HIE with four participating HIOs. During a visit to HIO1, the patient (William Smith) agrees to participate in the HIE and requests a VUPI from the HIE's identity provider on behalf of HIO1. The identity provider validates the request and issues a VUPI which is captured in HIO1's EMPI and associated with the patient's HIO1 electronic health record. In this example, HIO1 associates used the internal patient ID (i.e., 100) with the issued VUPI (i.e., 8562837139.7491742218) within their EMPI. The patient also receives an identification card from the identify provider who documents the issued VUPI. HIO1 also obtains the patient's consent to exchange all, some and none of his/her electronic health record with HIO2, HIO3 and HIO4, respectively. The patient subsequently provides his VUPI to HIO2, HIO3 and HIO4, who in turn capture it within their respective EMPIs and associates it with their respective electronic health record for the patient. HIO2, HIO3 and HIO4 are now able to unambiguously locate the patient's HIO1 electronic health record, determine which document(s) the patient has consented to share and retrieve all or some of those document(s) using VUPI-based patient-match, available-patient-documents and patient-document-retrieve queries, respectively. Figure 1 (next page) is provided to help illustrate this sample of the HIE and VUPI transactions.

Even though each EMPI captures different demographic traits for William Smith, the VUPI allows each HIO to uniquely identify the patient's electronic health record in one another's EHR. Statistical matching cannot deliver these results. For example, if HIO3's statistical matching algorithm heavily

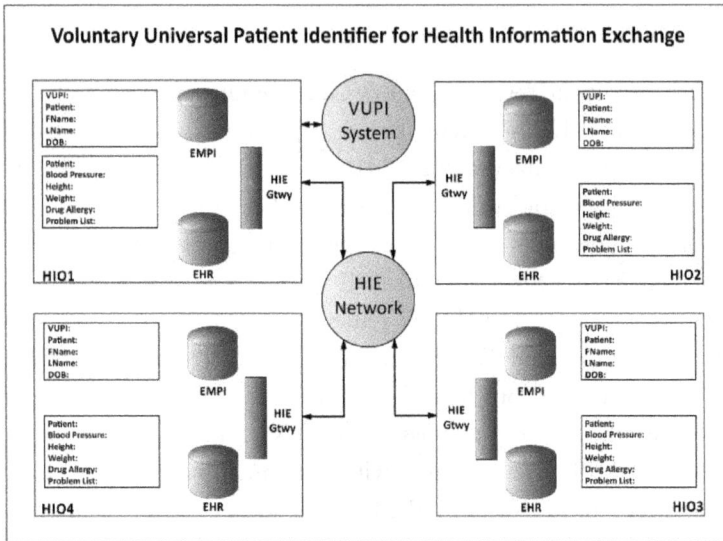

Figure 1: VUPI for HIE

weighted positive matching on a patient's social security number, then its patient-match query of HIO1 would return a false-negative result since HIO1 does not capture the patient's social security number. If HIO2's statistical matching algorithm heavily weighted positive matching on a patient's name and date of birth, then its patient-match query of HIO1 could return a false-positive result since HIO1 could have a different patient named William Smith born on July 11, 1965 (there are over 24,000 William Smiths in the United States (https://names.whitepages.com/william/smith, 2012)).

The risk of similar false-positive results has a higher probability in countries like Vietnam where about 40% of the population shares the same last name (http://english.vietnamnet.vn/en/society/19435/vietnam-s-nguyen-family-name-most-common-in-the-world.html).

## Conclusion

Few healthcare or insurance companies with access to personal health information give patients complete control over who can view and use information in their medical records.

Critics of UPIs cite concerns such as unauthorized use "for profit" and security risks that could result in medical identity theft, falsification of medical records, and the fraudulent use of information to obtain multiple UPIs. The issue is not whether National Patient Identifiers or Universal Patient Identifiers are beneficial. The issue is that not enough progress has been made in developing security systems powerful enough to prevent security breaches, offer oversight for how patient health information is used, and inform policies that provide patients with the right to control who has access to their personal health information and how this information is used. Once those problems are resolved, use of EHRs and UPIs will be highly beneficial in improving healthcare.

### Patient Matching Within a Health Information Exchange

*by Tim Godlove, PhD, and Adrian W. Ball, PMP, ITIL, CPHIMS*

American Health Information Management Association,
Perspectives in Health Information Management, Spring 2015

#### Abstract

The purpose of this article is to describe the false match/mismatch problems resulting from the Nationwide Health Information Network's automated patient discovery specification and propose a more effective and secure approach for patient matching between health information organizations participating in a health information exchange. This proposed approach would allow the patient to match his or her identity between a health information organization's electronic health records (EHRs) at the same time the patient identifies which EHR data he or she consents to share between organizations. The patient's EHR username/ password combination would be the credential used to establish and maintain health information exchange identity and consent data. The software developed to support this approach (e.g., an EHR health information exchange module) could also allow a patient to see what health information was shared when and with whom.

## Introduction

The electronic exchange of health information, commonly referred to as health information exchange (HIE), plays a vital and central role in delivering coordinated, accountable, patient-centered care that is both less expensive and of higher quality. HIE between two health information organizations (HIOs) has two key prerequisites. The first is confirming that the requesting and responding HIOs' patients are one and the same. The second is confirming that the patient has consented to the exchange of the health information between the HIOs. Many of today's HIEs, including the Virtual Lifetime Electronic Record (VLER) of the Department of Defense (DoD) and the Department of Veterans Affairs (VA), are implemented using the Nationwide Health Information Network (NwHIN) specifications. The NwHIN specifications describe an automated patient discovery process for confirming patient identity and a manual process for collecting patient consent. However, standardization issues with the data used by the NwHIN's automated patient discovery process commonly result in false mismatches and even false matches, both of which can cause major patient safety issues. Incorrectly mismatching or matching a patient as part of the HIE process may also have privacy and security implications. One way to resolve this issue to have patients confirm their HIO identity when they provide their consent to exchange data between HIOs.

As emerging new delivery systems move more aggressively toward sharing identifiable health information across disparate settings, concerns about historically suboptimal levels of accuracy in matching patients to their health information are exacerbated by poor data quality and incomplete data collection. The importance of the ability to link data is best put by Scheuren (1997): "Record linkage can aid a society in achieving advances in the well-being of its citizens."[1]

## Patient Matching

Patient matching is a relatively new process used by private and public healthcare institutions to detect common denominators in electronic health records (EHRs) on the basis of personal traits such as demographics, geographies, and histories. It is used to match patients with medical records when the records come from a variety of sources. That way, the patient can be identified by the records, instead of the other way around. The data may be retrieved from patient records at different hospitals, doctors' offices, and governmental databases. While patient matching applications have benefits such as the ability to identify best practices and to synthesize treatment, they also raise serious concerns about issues such as false matches, privacy rights, and patient consent.

As noted previously, two fundamental prerequisites for HIE between HIOs are confirming that the requesting and responding HIOs' patients are one and the same and confirming that the patient has consented to exchange the data. This prerequisite data must be stored and maintained within the HIE application. Figure 1 illustrates HIE between three hypothetical HIOs (Acme HIO, A1 HIO, and HIO-R-Us). The HIE system has confirmed that the patients in question are one and the same (i.e., John Doe in the Acme HIO [Patient ID 123] is John Doe in the A1 HIO [Patient ID 789] and is also John Doe in the HIO-R-Us HIO [Patient ID 456]) and has captured this patient match in the HIOs' HIE metadata stores. The HIE system has also confirmed what type of EHR data the patient has consented to exchange (e.g., John Doe has consented to exchange his lab results and prescriptions with HIO-R-Us and A1 HIO, respectively) and has confirmed that the HIOs have captured this consent information in their respective HIE metadata stores.

## Issues of Concern

The patient matching process is as controversial as any medical issue that involves potential invasions of privacy. It is not the matching process itself that is at the core of ethical concerns, but the overall ability to access the data. What makes this issue particularly sensitive is that in addition to the potential privacy violations, the process has a high probability of error. False positives, in which matches that should not have been made are made, and false negatives, in which matches that should have been are not made, are both dangerous and daunting. Until errors can be reduced and confidentiality (allowing access only to meet medical treatment needs and not for marketing purposes) and voluntary consent become a part of the process, patient matching will remain a controversial and volatile issue in the healthcare industry and among the general public.

Patient matching is a challenge for the healthcare industry as a whole and for HIOs in particular because of the need to ensure accuracy across organizations. The Office of the National Coordinator for Health Information Technology (ONC) has provided funding for a number of health information technology (IT) programs, including the development of the NwHIN, which is "a set of standards, services, and policies that enable the secure exchange of health information over the Internet."[2]

## Patient Discovery: The NwHIN Standard for Patient Matching

Patient Discovery is the NwHIN standard for HIE patient matching. Prior to exchanging patient-specific data, HIOs need to confirm that they are dealing with the same patient. The NwHIN patient discovery specification dictates how HIOs will locate and identify patient information that resides on another HIO on the NwHIN.

The North Carolina Healthcare Information and Communications Alliance describes the NwHIN patient discovery process as follows: "The initiating HIO enters all the demographic data and local identifiers that can be shared about a patient. The responding HIO matches the demographics and identifiers. If a single match is found that is considered highly reliable, it is returned to the initiator, along with its demographic details and identifiers. If no patient match is found then the responder sends an empty response to the initiator, indicating that this patient is not known at this HIO. If a highly reliable match cannot be identified, an ambiguous response is returned."[3] At this point, the NwHIN patient discovery specification requires additional information or the use of a manual process to finish the patient matching process. The NwHIN patient discovery specification "is designed to avoid false matches at all costs."[4]

The NwHIN patient discovery specification dictates that patient demographic traits be exchanged between HIOs and that each HIO is to respond with a match or an empty reply. When a match is found, the responder sends back the patient's demographic traits and patient ID. In a study of a NwHIN-based HIE, Bouhaddou et al. (2012) note that this process "enables the sender to validate the match using the demographics received" and "is considered a 'no risk' approach. However, the patient match success rate is far less than ideal."[5] Bouhaddou et al. note that as of May 9, 2012, only 23,611, or 53 percent, of veterans who consented to share their data were successfully matched by the VA's NwHIN-based HIE with the DoD and several private-sector HIOs and that "the failures were mainly due to lack of accurate, standardized, complete data. In other words, without a complete set of the primary identifiers (i.e., first name, last name, middle name/initial, gender, date of birth, and social security number), it is unlikely to match a patient. Other reasons include partial coverage of opted in Veterans population, requirement for a second authorization, and difference in patient matching algorithms."[6]

Bouhaddou et al. further note that "other insights related to Identity, Privacy, and Consent Management gained from [this NwHIN-based HIE] include" the following:[7]

1. "When onboarding a new [HIO] to NwHIN that has millions of patients, creating initial [patient matches] is challenging. [Patient discovery] as a broadcast-out model is difficult to scale for a future NwHIN that supports hundreds of HIEs."[8]

2. The NwHIN patient discovery "specifications do not provide clear guidance on how to keep the patient [matches] up-to-date when there are patient ID changes (e.g., marriage), merges, etc."[9]

3. "Inclusion of the full Social Security Number (SSN) in the demographic traits is necessary for any reasonable level of matching success. Some organizations or states do not exchange SSNs or only exchange the last four digits, which makes it unlikely a unique match can be achieved."[10]

4. "Lack of a 'common/ standard' consent model is a barrier. Patients may have to sign multiple consents, one for each HIO, before their data are shared."[11]

5. "Specification would benefit from more clear definitions of the ...attributes [of the data to be exchanged] and more complete . . . samples of the permitted values."[12]

### The Risks Associated with Exchanging Health Data on the Basis of a False Match

The major risk involved in using automated deterministic or probabilistic patient matching algorithms like the NwHIN's patient discovery is the occurrence of false matches. According to the Bipartisan Policy Center, "Error rates, which average eight percent and can range up to 20 percent—can result in sub-optimal care and medical errors," and "incorrectly matching a patient to a health record may also have privacy and security implications, such as wrongful disclosure—in addition to treatment based on another patient's health information."[13]

The two main reasons for false matches are human input error (e.g., misspellings, number reversals) and lack of standardization. Human error is natural, but it can be reduced through improvement of the other area contributing to false matches, that is, standardization.[14] According to the Bipartisan Policy Center, "The lack of standardization in the data attributes or fields used for matching, the information contained in those fields, and methods used, results in increased error rates as well as significant burden and cost within the health care system."[15]

Data exchange capabilities are a critical issue in developing national standardization. Developing compatible software and hardware that interface with one another on a national level is a monumental undertaking in terms of both logistics and cooperation. According to Stead et al. (2005), "Since the United States does not have a national patient identifier, data interchange begins with a mechanism—such as a master patient index or its equivalent function—to determine which records relate to a single person. A mechanism is needed to authorize access to data by a person at a site other than the place it originated. Such a service, therefore, is as much about governance, trust, and common terminology as it is about technology."[16]

Synthesizing regulations and standards is as big of a challenge as is synchronizing software and hardware. A major part of the problem stems from a lack of compliance with regulation. A 2010 Identity Force survey, involving more than 200 compliance experts across the United States, revealed that, despite increases in laws and regulations relating to compliance with security standards, the number of security violation incidents in the healthcare industry continues to rise.[17]

In 2013, the ONC released the findings of an extensive study on patient matching that involved "more than 50 large health systems and health IT software developers."[18] The study had two primary objectives: "to define common features that achieve high positive match rates across different systems; and to define the processes and best practices that are most effective to support high matching rates."[19] The recommendations that were derived from the findings of the study centered on standardization. In particular, the ONC recommended that the following information be standardized in every HIE transaction:

- Current and past addresses
- Date of birth
- Full name
- Gender
- Phone numbers[20]

Other recommendations made by the ONC in its report included the following:

- "Additional data attributes to improve patient matching should be studied."
- "An open source algorithm should be developed to test and build patient matching capabilities."
- "Certified electronic health record systems should be required to generate and provide reports that detail possible duplicate patient records."
- "EHR certification criteria should include the ability to capture patient identifying attributes."[21]

The report also recommended development of the following:

- "Formal best practices for patient matching and data governance."
- "Policies to encourage consumers to keep their health information accurate and up to date."
- Educational and training materials for verifying patient data attributes."[22]

Most hospitals and other healthcare facilities are in need of more streamlined methods of communication and data sharing with other facilities and government agencies. Progress is being made, but a strategically designed integration strategy in which systems and interactions are standardized is still needed. Standardization will increase cost efficiency and decrease the potential for errors. However, it will not do much to help with the second major area of concern regarding patient matching, which is privacy.

## Privacy Issues Related to National Patient Identifier–based Patient Matching

The main objective of patient matching is to provide more synergistic, comprehensive treatment and care to patients, but a secondary goal is to exchange and sell patient information for marketing endeavors that lead to monetary gain. This secondary goal is one of the primary risks associated with patient matching because it destroys expectations of privacy and security. Most patients do not mind other doctors' being able to view their records for medical treatment purposes, but the idea of drug companies viewing the records, just so they can send advertisements, does not sit well with many people. According to Peel (2013), "'Patient matching' is a method of involuntary, hidden surveillance, much like the NSA's surveillance of phone records and metadata. It enables 1000s of hidden third parties to collect and aggregate our personal health data from many places without our knowledge or consent."[23]

The increased use of EHRs has given rise to a host of concerns regarding patient privacy and the confidentiality of patient data. One proposed, yet controversial solution to the problems associated with electronic health data exchange is the Unique Patient Identifier (UPI). The Health Insurance Portability and Accountability Act of 1996 (HIPAA) supported the development of a UPI. However, in 1999, Congress took countermeasures to prevent the creation of UPIs because of widespread concerns about privacy and confidentiality. Congress passed a law that prohibited the US Department of Health and Human Services from allocating any of its budget toward UPI development unless Congress approved it beforehand. Fifteen years later, this restriction is still active.[24]

Public perceptions of UPI development vary, but almost everyone agrees on a couple of points. The first is that although UPIs would be helpful, they will not be a cure-all for the issues that plague patient matching. The second is that patients should have to sign a consent form allowing for an identifier to be used only with their express permission. To make UPI use mandatory without gaining consent from patients brings to light numerous potential ethical violations, and possibly even legal ones.[25]

This understanding has led to proposals for a voluntary private-sector option known as the Voluntary Universal Healthcare Identifier (VUHID). This voluntary approach to assigning patients their own unique ID numbers eclipses the majority of objections that people have vocalized regarding mandatory national patient identifiers. According to Paxton (2009), "VUHID is based on two standards developed by ASTM (originally the American Society of Testing and Materials) and ANSI (American National Standards Institute). The system's goal is to make unique health care identifiers available at nominal cost to individuals who want one. But more than that, VUHID promises to shield patient privacy by using two categories of the identifier: an open identifier for information a person wants to have known to all of his or her care providers, and multiple private identifiers for medical information a person wants to keep private."[26] This solution seems to offer an ideal compromise. However, there is still a fair amount of objection to the concept, perhaps based on the principle alone.

While a blanket solution does not seem to be available, the Bipartisan Policy Center (2012) recommends that the following measures should be taken to ensure the accuracy, efficiency, and ethicality of patient matching applications such as UPIs:[27]

1.      "Standardize Matching Methods... This includes standardizing data fields, definitions and validation methods designed to improve the accuracy and the quality of the information gathered from patients."[28]

2.      "Standardize Policies... including those related to the establishment of acceptable benchmarks or rates of error in matching."[29]

3.      "Share Lessons Learned and Best Practices. . . regarding technology, human resources, workflow and policy. . . . More transparency in disclosing accuracy rates will facilitate assessment of methods and also promote improvement."[30]

4.      "Collectively Organize and Support the Adoption of Shared Services. Common principles, policies, standards, and methods for matching patient data will facilitate the sharing of services for matching across many organizations, promoting standardization, improving results and producing economies of scale."[31]

Consent is without a doubt one of the biggest issues related to patient matching applications, in terms of both practical issues and ethical concerns. In general, people do not favor the concept of strangers being able to look at their private medical data without consent or strangers using or distributing private medical data however they please, without the patient's knowledge or consent. However, the voluntary model has failed to gain momentum, primarily because interested parties are convinced that most people will staunchly refuse to provide their consent.

### The Need to Collect Patient Consent as Part of the Matching Process

Consent, as part of the patient matching process, encompasses issues related to permission, access, and transparency. According to Peel (2013), "Today, the nation's sensitive health records are exchanged by hundreds of hidden users without meaningful informed consent. Health technology systems violate our federal rights to see who used our data and why. Despite the federal right to an Accounting of Disclosures (AODs)—the lists of who accessed our health data and why—technology systems violate this right to accountability and transparency."[32]

For people to accept the idea of patient matching and national patient identifiers, they need to know that their involvement in the process matters. They need to know that nothing can be done without their express permission. They need to feel empowered in the decision-making process regarding who is allowed to access their personal medical information and who is not. They need to know that they will have access to the matching data so that they can personally verify whether the data is accurate or is in error. Until these needs are fulfilled, the national acceptance of UPIs is going to remain highly unlikely.

### Solution

To address this problem, the following solution is proposed: Have the patient validate his or her identity within the HIE's

HIOs and collect the patient's HIE consent information as part of the patient matching process. Figure 2 illustrates HIE between three HIOs (Acme HIO, A1 HIO, and HIO-R-Us). The HIE application would collect a patient's Acme HIO, A1 HIO, and HIO-R-Us login credentials (username and password) and then validate those credentials to establish a patient match between these HIOs. As part of this process, the patient would also identify what EHR information he or she consents to share as part of the HIE (e.g., John Doe has consented to share his Acme HIO EHR lab results and prescriptions with HIO-R-Us and A1 HIO, respectively). As in the model described in Figure 1, this patient match and consent data is captured in the HIE metadata stores. This patient matching and consent approach assumes that patients have login credentials for the HIOs to which they belong.

## Conclusion

As noted previously, the patient matching process is a controversial problem that involves possible invasions of privacy and a high probability of error. The NwHIN's automated patient discovery specification has the potential for false match/mismatch errors that may result in patient safety problems and privacy and security implications.

The alternative approach described in this article would allow the patient to match his or her identity between one HIO's EHR at the same time that the patient identifies which EHR data he or she consents to share between HIOs. The patient's EHR username/password combination would be the credential used to establish and maintain HIE identity and consent data. The software developed to support this approach (e.g., an EHR HIE module) could also allow a patient to see what health information was shared when and with whom.

Until such an approach can be implemented and confidentiality and voluntary consent become an integral part of the process, patient matching will continue to be a source of concern both within the healthcare industry and among the general public.

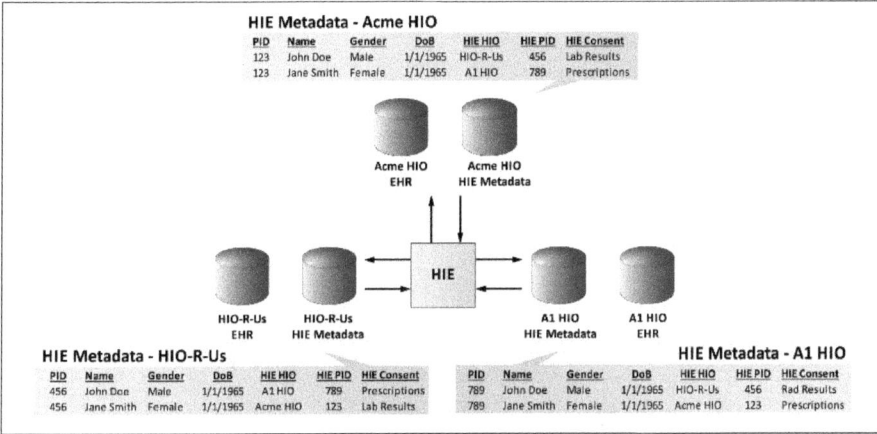

Figure 1: HIE Metadata - Acme HIO

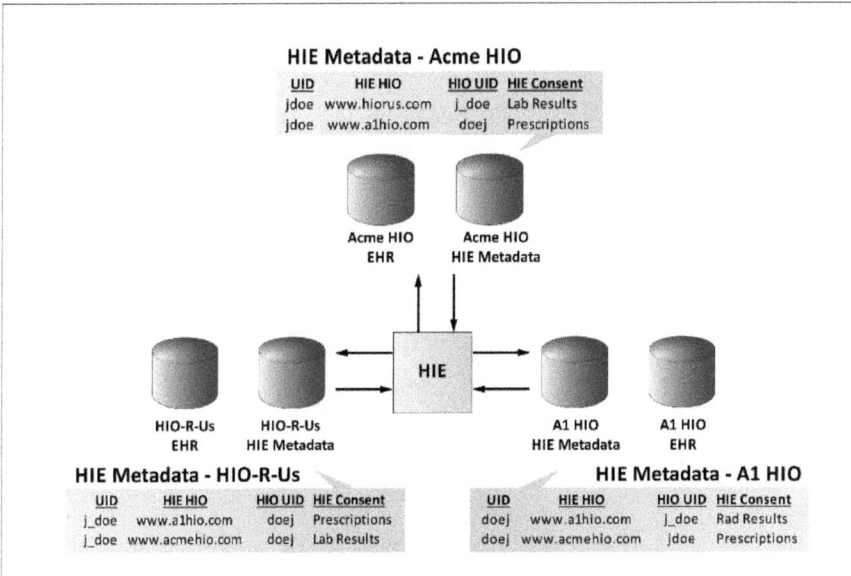

Figure 2: HIE Metadata - Acme HIO

# References

All online resources accessed and verified early 2015, but subject to changed and deletion.

**Notes to Chapter 1**

Graziano, C. (2003, September) Learning to live with biometrics. Wired. Retrieved from http://www.wired.com/politics/security/news/2003/09/60342

Halboob, W., Mahmod, R., Udzir, N. I., & Abdullah, M. T. (2015). Privacy policies for computer forensics. *Computer Fraud & Security, 2015*(8), 9-13.

Hosmer, C., Gordon, G., Hyde, C., Grant, T. Cyber Forensics 2000. Proceedings, 1st Annual Study of the State-of-the-Art in Cyber Forensics

Markoff, J. (2010, July 4) Do we need an internet identity card? New York Times, p. 3

Marx, G.T. (2001) Identity and anonymity: Some conceptual distinctions and issues for research. In J. Caplan and J. Torpey, (2001) Documenting Individual Identity. Princeton University Press.

National Identification Cards: Why Does the ACLU Oppose a National I.D. System? (2015), ACLU. Retrieved from https://www.aclu.org/national-identification-cards-why-does-aclu-oppose-national-id-system

Northcutt, S., Zeltser, L., Winters, S., Kent, K., & Ritchey, R. W. (2005). *Inside Network Perimeter Security*. Sams.

Pasley, J.F. (2003) United States Homeland Security in the Information Age: Dealing with the Threat of Cyberterrorism, *White House Studies, 3*, 403

Poulsen, K. (2008, May 29) Comcast hijacker say they warned the company first, Wired, Retrieved from http://www.wired.com/2008/05/comcast-hijacke/

Poulsen, K. (2009) Three charged as Comcast hackers. Wired, Retrieved from
http://www.cnn.com/2009/TECH/11/20/comcast.hacking.charge/

Rawlins G. J. E. (1996). Moths to the Flame: The Seductions of Computer Technology. Cambridge, Massachusetts and London: MIT Press.

Rosenblatt, H. J. (2014). *Systems analysis and design.* Cengage Learning.

Thalheim, B. (2013). *Entity-relationship modeling: Foundations of database technology.* Springer Science & Business Media.

Turner, A. (2010) Biometrics: Applying an emerging technology to jails, *Corrections Today,* (2000) 62(6), 26-27.

Vacca, J. R. (2007) Biometric technologies and verification systems. Butterworth-Heinemann.

Vatis, M.A., (2000) *Statement of Michael A. Vatis, Director, National Infrastructure Protection Center Federal Bureau of Investigation before the Senate Committee on Judiciary,* May 25, Retrieved from http://www.steptoe.com/assets/attachments/1012.htm

**Notes to Chapter 2**

"Biometric access control, A good choice for homeowners?" (2006) www.homesecurityinformation.com

Clarke, N, L. & Furnell, S. (2007) Authenticating mobile phone users using keystroke analysis. *International Journal of Information Security,* 6(1), 1-14.

"Defending anonymity" (2008) Anarchist Federation, www.afed.org.uk

Graziano, C. (2003, September) Learning to live with biometrics. *Wired.* Retrieved from http://www.wired.com/politics/security/news/2003/09/60342

Markoff, J. (2010, July 4) Do we need an internet identity card? New York Times, p. 3

Marx, G.T. (2001) Identity and anonymity: Some conceptual

distinctions and issues for research. In J. Caplan and J. Torpey, (2001) *Documenting Individual Identity*. Princeton University Press.

National Identification Cards: Why Does the ACLU Oppose a National I.D. System? (2002, March 12), ACLU. Retrieved from http://www.aclu.org/immigrants/gen/11666res20020312.html

Turner, A. (2000) Biometrics: Applying an emerging technology to jails, *Corrections Today*, (2000) 62(6), 26-27.

Vacca, J. R. (2007) *Biometric technologies and verification systems*. Butterworth-Heinemann.

*Wilson, T.V. (2010) How* biometrics works. Howstuffworks.com.

### Notes to Chapter 3

Charette, R. (2006, June) EHRs: Electronic Health Records or Exceptional Hidden Risks? *Communications of the ACM, 49*

Duncan, G.T. & Mukherjee, S. (2000) Optimal disclosure limitation strategy in statistical databases: deterring tracker attacks through additive noise, *Journal of the American Statistical Association*, 95, 720-731

Fetter M. S. (2009) Electronic health records and privacy. *Issues in Mental Health Nursing. 30:* 408-409

Fredericks, P. (2007, September 1) The corporate culture perspective, Security Magazine online, http://www.securitymagazine.com/CDA/Articles/Feature_Article/BN P_GUID_9-5-2006_A_10000000000000159632

Healthcare Technology Management (2009, February) Privacy measures for EHRs.

Kandra, A. (2004, December) Trusting Your Health History to the Web Digitized medical records can help save lives. But are they secure enough? *PC World*

Lowes, R. (2006, March 17) Healthcare IT: How safe is your patient data? *Medical Economics*.

Maheu, M. M., Pulier, M. L., Wilhelm, F. H., McMenamin, J., & Brown-Connolly, N. (2005). *The mental health professional and the new technologies: A handbook for practice today.* Mahwah, NJ: Lawrence Erlbaum Associates

National Committee on Vital and Health Statistics (NCVHS) (2007). Enhancing Protections for Uses of Health Data: A Stewardship Framework, U.S. Department of Health and Human Services

Pector, E.A. (2009, July 10) Top-down, bottom-up, and medicine in the middle: Reform efforts are barreling forward with physicians tied to the tracks, *Medical Economics, 86*, 43-45

**Notes to Chapter 4**

Alfreds, S.T., Tutty, M., Savageau, J.A., Young, S. & Himmelstein, J. (2006) Clinical Health Information Technologies and the Role of Medicaid. *Health Care Financing Review. 28* (2). 11-15

Andrews, M. (2008, February 29) Medical identity theft turns patients into victims. U.S. News and World Reports. Retrieved from http://health.usnews.com/health-news/family-health/articles/2008/02/29/medical-identity-theft-turns-patients-into-victims.html

Eng, T. (2001). The ehealth landscape: A terrain map of emerging information and communication technologies in health and health care. Robert Wood Johnson Foundation. Retrieved from http://www.informatics-review.com/thoughts/rwjf.html

Fay, J. (2005) *Contemporary security management*, (2d ed) Butterworth-Heinemann

Goedert, J. (2010, May 1) OCR shines a harsh light on data breaches, *Health Data Management 18*(5).26.

Identity Force (2010, April 20) Delayed compliance with New Regulations Has Increased Data Breaches and Medical Identity Theft in U.S. Hospitals. Retrieved from http://aha-solutions.org/aha-solutions/content/HR/IdentityForce/HospitalSurvey-PR-4-20-10final.pdf

Koleva, G. (2010, June 4) Medical ID theft a threat to privacy, health and finances, Walletpop.com, Retrieved from http://www.walletpop.com/blog/2010/06/04/increasing-medical-id-theft-a-threat-to-privacy-health-and-fina/

McGraw, D. (2010, June 12) California hospitals fined as medical records are breached. The Sacramento Bee. Retrieved from http://www.cdt.org/press_hit/california-hospitals-fined-medical-records-are-breached

Prince, K. (2008, October 28) *Health care data security breaches in the U.S.* Perimeter eSecurity

Sanchez-Abril, P, & Cava, A. (2008). Health privacy in a techno-social world: a cyber-patient's bill of rights. *Northwestern Journal of Technology and Intellectual Property*, (6), 244-277.

**Notes to Chapter 5**

Contos, B. (2007, April 1) It's collaboration time, Security Magazine online http://www.securitymagazine.com/CDA/Archives/BNP_GUID_9-5-2006_A_10000000000000077101

Fredericks, P. (2007, September 1) The corporate culture perspective, Security Magazine online, http://www.securitymagazine.com/CDA/Articles/Feature_Article/BNP_GUID_9-5-2006_A_10000000000000159632

Kilpatrick, I (2007, November). Dam data leakage at source: how unified encryption management (UEM) is changing the threat landscape, *Software World*, 12(4).

Kouzes, J. M., & Posner, B. Z. (1995). *The leadership challenge: How to get extraordinary things done in organizations.* San Francisco: Jossey-Bass.

National Institute of Standards and Technology. *"Guide to Selecting Information Technology Security Products"*, SP-800-36. October 2003

National Strategy to Secure Cyberspace (2003, February) http://www.dhs.gov/xlibrary/assets/National_Cyberspace_Strategy.pdf

Wade, J.R. (2004, May 4) Compliance made easy, SC Magazine online http://www.scmagazineus.com/

**Notes to Chapter 6**

Antonopoulos, A.M. (2007). Combining work and play threatens business security. *Network    World, October 10*, 1-2

Auten, C. (2007). Ready, set, go: hot to put telework in the fast lane. *EWeek*, 1-4.

Axelson, C., Wardh, I., Strender, L.E. & Nilsson, G. (2007). Using medical knowledge sources on handheld computers—a qualitative study among junior doctors. *Medical Teacher, 29*, 611-622.

Edwards, C. (2005). Wherever you go, you're on the job. *BusinessWeek, June 20*, 1-5.

Friedman, J. & Hoffman, D.V. (2008). Protecting data on mobile devices: a taxonomy of security threats to mobile computing and review of applicable defenses. *Information Knowledge Systems Management, 7*, 159-180.

Jackson, B. (2008). SONY site falls prey to automated hacker hijack. *PCWorld, September* 1, 56.

Kaven, O. (2004). Keep your office safe. *PC Magazine,* www.pcmag.com, August 3, 93-101.

Knorr, E. (2004). Is true mobility at hand? *Infoworld, May 10*, 33-34.

Wellman, B., Salaff, J., Dimitrova, D., Garton, L., Gulia, M. & Haythornthwaite, C. (1996). Computer networks as social networks: collaborative work, telework and virtual community. *Annu, Rev. Sociol, 22*, 213-238.

**Notes to Chapter 7**

Antonopoulos, A. M. (2007, October 10) Combining work and play threatens business security. *Network World,* http://www.networkworld.com/columnists/2007/101007-risk-reward.html?fsrc=rss-antonopoulos

Auten, C. (2008, March 3) Ready, set, go: How to put telework in the fast lane, eWeek.com, http://www.eweek.com/c/a/Enterprise-Applications/Ready-Set-Go-How-to-Put-Telework-in-the-Fast-Lane/

Ajzen, I. (2005) *Attitudes, personality and behavior*, Maidenhead, England: Open University Press

Bain, B. (2007, September 13) Justice says no to private PCs for telework, FCW.com, http://www.fcw.com/online/news/150044-1.html

Flood, K. (2001, February 5). The forgotten side of network security. *Network World, 12*, 276-283.

Hines, M. (2007, August 21) Mobile workers still struggling with security; A new study shows that even as the business use of mobile devices increases, many users are unconcerned or uninformed about security issues and practices." *InfoWorld*, http://www.infoworld.com/article/07/08/21/Mobile-workers-still-struggling-with-security_1.html

Jones, K.C. (2007, November 5) Businesses more concerned about mobile, remote security, but still ignore training, *InformationWeek*, http://www.informationweek.com/news/showArticle.jhtml?articleID=202802456

Kilpatrick, I (2007, November). Dam data leakage at source: how unified encryption management (UEM) is changing the threat landscape, *Software World*, 12(4).

Lampson, B. (2005). Microsoft, Accountability and Freedom, http://research.microsoft.com/lampson/slides/accountabilityAndFreedomAbstract.htm.

Mears, J. (2007, April) Legislation promotes Federal teleworkers; Telework Enhancement Act of 2007 would open more doors to telecommuting. *Network World*, http://www.nwwsubscribe.com/news/2007/040307-telework-legislation.html

National Institute of Standards & Technology. (2007). *SP 800-114, User's Guide to Securing        External Devices for Telework and Remote Access*: National Institute of Standards and        Technology.

Sternstein, A. (2007, June 4) Survey: Unauthorized teleworkers a security risk, *National Journal's Technology Daily*, http://www.govexec.com/story_page.cfm?articleid=37105

**Notes to Chapter 8**

Berman, J. & Bruening, P. (2007). *Is privacy still possible in the twenty-first century?* Washington, DC: Center for Democracy & Technology. Retrieved from http://old.cdt.org/publications/privacystill.shtml

Lewis, J. A. (2011). Cybersecurity two years later: A report of the CSIS commission on cybersecurity for the 44th presidency. Washington, DC: Center for Strategic and International Studies. Retrieved from http://csis.org/files/publication/110128_Lewis_ CybersecurityTwoYearsLater_Web.pdf

Lohr, S. (2011, Jan. 1). Computers that see you and keep watch over you. *The New York Times.* Retrieved from http://www.nytimes.com/2011/01/02/science/02see.html

McCullagh, C. (2011). Senator renews pledge to update digital privacy law. *CNET News.* Retrieved from http://news.cnet.com/8301-31921_3-20071670-281/senator-renews-pledge-to-update-digital-privacy-law/#ixzz1RljkbnBg

**Notes to Chapter 9**

Andress, A. (2003) *Surviving Security: How to Integrate People, Process, and Technology.* Boca Raton, FL: Auerbach.

Bielski, L. (2004). Keep Proprietary Information in Its Place: Running an 'Enterprise' with        Control, Insight, and an Eye on Risk Management. *ABA Banking Journal,* 96(4): 52-58.   Retrieved May 5, 2008, http://www.allbusiness.com/legal/laws/137657-1.html

Collin, B. (1997). The Future of Cyberterrorism. *Crime and Justice International,* March    1997:5–18. Retrieved May 1, 2008, http://www.cjcenter.org/cjcenter/publications/cji/archives/cji.php?id =415

Guttman, R. (2003). *Cybercash: The Coming Era of Electronic Money.* New York: Palgrave   McMillan.

Lineberry, S. (2007) "The Human Element: The Weakest Link in Information Society." *Journal       of Accountancy,* 204(5):44 –50. Retrieved May 1, 2008, http://www.aicpa.org/pubs/jofa/nov2007/human_element.htm

Sharma, A. (2006) " A Holistic Approach to Physical and IT Security." *Security Magazine.* 21     November. Retrieved April 28, 2008, http://www.securitymagazine.com/CDA/Articles/Feature_Article/c80

5be157db0f010VgnVCM100000f932a8c0

## Notes to Chapter 10

Cavoukian, A. (2006). 7 laws of identity: Privacy-imbedded identity in the digital age. Retrieved from http://www.identityblog.com/wp-content/resources/7_laws_whitepaper.pdf

Cohen, S. S., DeLong, B., & Weber, S. (2001). Tools: The drivers of e-commerce. In The BRIE-IGCC economy project (Ed.), *Tracking a transformation: E-commerce and the terms of competition in industries.* Washington, DC: Brookings Press.

Holtzman, D. H. (2006). *Privacy lost: How technology is endangering your privacy.* Hoboken, NJ: Jossey-Bass.

Ladine, B. (2002, Aug. 10). Medical privacy rules are relaxed: New White House rule drops written consent. *The Boston Globe*, A1.

Markoff, J. (2002, Nov. 9). Pentagon plans a computer system that would peek at personal data of Americans. *The New York Times.*

Reporters without boarders press release. (2002, Sept. 5). *The Internet on probation: Anti-terrorism drive threatens Internet freedoms worldwide.*

Schwartz, P. M., & Reidenberg, J. R. (1996). *Data privacy law: A study of United States data protection.* Charlottesville, NC: Michie.

Shapiro, C., & Varian, H. R. (1999). *Information rules: A strategic guide to the network economy.* Boston: Harvard Business School Press.

Singletary, M. (2000, March 19). Credit rating: We should know the score. *The Washington Post*, H1.

Solove, D. J. (2004). *The digital person: Technology and privacy in the information age.* New York: New York University Press. Retrieved from http:docs.law.gwu.edu/faceweb/ dsolove/Digital-Person/text.htm

Thompson, J. F. (2002). Identity, privacy, and information technology. *Educause Review*, 64-65.

Waldo, J., Lin, H.S., & Millett, L. I. (Eds.). (2010). Engaging privacy and information technology in a digital age. *Journal of Privacy and Confidentiality, 2*(1), 15-18.

## Notes to Chapter 12

Badzek, L., Turner, M., & Jenkins, J. F. (2008). Genomics and nursing practice: Advancing the nursing profession. OJIN:The Online Journal of Issues in Nursing, 13(1).

Charette, R. (2006, ) EHRs: Electronic health records or exceptional hidden risks? *Communications of the ACM, 49*(6),120.

Consensus Panel, June 2008. (2009). Essentials of genetic and genomic nursing: Competencies, curricula guidelines, andoutcome indicators (2nd Edition). Silver Spring, MD: American Nurses Association.

Genetic Alliance. (2008, November). What does GINA mean? A guide to the genetic information nondiscrimination act. Retrieved from www.qeneticalliance.org/GINAResource

Genetic Alliance. (2009). Genetic discrimination. Retrieved from www.qeneticalliance.orq/policy.discrimination

Genetics and Public Policy Center. (2008a). Center launches new GINA resource. Retrieved from www.dnapolicy.org/news.release. php?action=detail&pressrelease id= 101

Genetics and Public Policy Center. (2008b). Frequently asked questions. Retrieved from www.dnapolicy.org/gina/faqs. html#insurance2

Genetics and Public Policy Center. (2008c). Genetic privacy and discrimination. Retrieved from www.dnapolicy.org/policy.privacy.php

Genetics and Public Policy Center. (2008d). Information on the genetic information nondiscrimination act (GINA). Retrieved from http://www.dnapolicy.org/resources/WhatGINAdoesanddoesnotdoch art.pdf

Hudson, K.L., Holohan, M. K., & Collins, F. S. (2008). Keeping pace with the times - The genetic informationn ond iscrimination act of 2008. The New England Journal of Medicine, 358 (25), 2661-2663.

Hellman, G. (2010, February 1) How to Secure Health Care Data to Meet HITECH Act Compliance. e.Week. Retrieved from http://www.eweek.com/c/a/Health-Care-IT/How-to-Secure-

Healthcare-Data-to-Meet-HITECH-Act-Compliance/

Identity Force (2010, April 20) Delayed compliance with New Regulations Has Increased Data Breaches and Medical Identity Theft in U.S. Hospitals. Retrieved from http://aha-solutions.org/aha-solutions/content/HR/IdentityForce/HospitalSurvey-PR-4-20-10final.pdf

Junglen, L. M., Pestka, E. L., Clawson, M. L., & Fisher, S.D. (2008). Incorporating genetics and genomics into nursingpractice: A demonstration. *OJIN: The Online Journal of Issues in Nursing*, 13(3).

Laedtke, A. L., O'Neill, S. M., Rubinstein, W. S., & Vogel, K. J. (2012). Family physicians' awareness and knowledge of the Genetic Information Non-Discrimination Act (GINA). *Journal of Genetic Counseling*, 21(2), 345-352.

Matloff, E. T., Shappell, H., Brierley, K., Bernhard, B. A., McKinnon, W., & Peshkin, B. N. (2000). What would you do? Specialists' perspectives on cancer genetic testing, prophylactic surgery, and insurance discrimination. *Journal of Clinical Oncology*, 18(12), 2484-2492.

McGraw, D. (2010, June 12) California hospitals fined as medical records are breached. The Sacramento Bee. Retrieved from http://www.cdt.org/press_hit/california-hospitals-fined-medical-records-are-breached

Monsen, R. B., (Ed.). (2009). Genetics and ethics in healthcare: New questions in the age of genomic health. Silver Spring,MD: Nursebooks.org

National Genetics Education and Development Centre. (2008). Telling Stories: Understanding real life genetics. Retrieved from www.geneticseducation.nhs.uk/tellingstories/

National Human Genome Research Institute. (2008a). Frequently asked questions about genetic testing. Retrieved from www.genome.gov/19516567.

National Human Genome Research Institute. (2008b). Genetic discrimination. Retrieved from www.qenome.qov/10002077

National Human Genome Research Institute. (2008c). Genomics in action: Getting ahead of the curve on genetic tests. Retrieved from www.genome.gov/15015000.

Stead, W.W., Kelly, B.J., & Kolodner, R.M. (2005). Achievable steps toward building a national health information infrastructure in the United States. Journal of the American Medical Information Association, 12(2), 113-120. Retrieved from http://www.ncbi.nlm.nih.gov/pmc/articles/PMC551543/

U. S. Department of Health and Human Services. (2008). Office of civil rights - HIPAA. Retrieved from www.hhs.gov/ocr/hipaa

**Notes to Chapter 13**

Arquilla, J. & Ronfedlt, D. (2001). Networks and Netwars: The Future of Terror, Crime, and Militancy. Wired. Retrieved from http://archive.wired.com/wired/archive/9.12/netwar_pr.html

Another Pearl Harbor Possible – Dreams of the Great Earth Changes. (n.d.). Retrieved from http://greatdreams.com/war/pearl-harbor-three.htm

Bell, D. & Renner, M. (n.d.). A New Marshall Plan? Advancing Human Security and Controlling Terrorism. Worldwatch Institute. Retrieved from http://www.worldwatch.org/new-marshall-plan-advancing-human-security-and-control

Bioterrorism Information and Resources. (n.d.). American Association of Nurse Anesthetists. Retrieved from http://www.aana.com/resources2/Pages/Bioterrorism-Information-and-Resources-Links-.aspx

Bush, GW (2001) Executive Order on Critical Infrastructure Protection. Retrived from http://www.whitehouse.gov/news/releases/2001/10/20011016-12.html

Cetron, Marvin J. (1989). The Growing Threat of Terrorism. The Futurist. Retrieved from https://www.questia.com/magazine/1G1-7698835/the-growing-threat-of-terrorism

Jennings, L., & Wagner, C. (2001). Cetron Interview - World Future Society. Retrieved from http://www.wfs.org/intcetron.htm

Defending America in the 21st Century. (2000) Retrieved from
http://csis.org/files/media/csis/pubs/defendamer21stexecsumm.pdf

Full Alert, RAND. (2001). Retrieved from
http://www.rand.org/pubs/periodicals/rand-review/issues/rr-12-01/fullalert.html

Morton, O. (September 2009). Divided We Stand. Wired. Retrieved from http://archive.wired.com/wired/archive/9.12/defense_pr.html

Opinion: An authoritarian agenda. (September 29, 2001). Tampa Bay Times. Retrieved from
http://www.sptimes.com/News/092901/Opinion/An_authoritarian_a gen.shtml

Schweitzer, Glenn E., Dorsch, Carole C. (1999). Superterrorism: Searching for Long-Term Solutions. The Futurist. Retrieved from https://www.questia.com/magazine/1G1-54772930/superterrorism-searching-for-long-

The Department of Defense Critical Infrastructure Protection (CIP) Plan. (November 18, 1998). Retrieved from
http://fas.org/irp/offdocs/pdd/DOD-CIP-Plan.htm

### Notes to Chapter 14

Baker, Stephen (2011). National Strategy for Trusted Identities in Cyberspace, Enhancing Online Choice, Efficiency, Security, and Privacy.

Brown, C. L. (2012). Health-care data protection and biometric authentication policies: Comparative culture and technology acceptance in China and in the United States. *Review of Policy research, 29*(1), 141-159. Retrieved from
http://econpapers.repec.org/article/blarevpol/v_3a29_3ay_3a2012_3ai _3a1_3ap_3a141-159.htm

Carter, B. (2011). Electronic Medical Records: A Prescription for Increased Medical Malpractice Liability? *Vanderbilt Journal of Entertainment and Technology Law 13*(2), 385-406. Retrieved from http://www.jetlaw.org/wp-content/uploads/2011/ 05/Carter-FINAL.pdf

Codd, E. (1985). "Is Your DBMS Really Relational?" and "Does

Your DBMS Run By the Rules?" ComputerWorld, October 14 and October 21.

Dimitropoulos, Linda L., Ph.D. (2009). Privacy and Security Solutions for Interoperable Health Information Exchange – Perspectives on Patient Matching: Approaches, Findings and Challenges. Retrieved from *healthit.hhs.gov/...0.../PatientMatchingWhite_Paper_Final.pdf*

U.S. Department of Health and Human Services. (1998). Unique health identifier for Individuals: A white paper. Retrieved from http://epic.org/privacy/medical/hhs-id-798.html

Healthcare Information Management and Systems Society (HIMSS). (2011). *Nationwide patient identification solution issue* [Online]. Retrieved from http://www.himss.org/policy/d/HGRR/201110_NationwidePatient_ID_SolutionPresentation.pdf

Hieb, Barry, MD, Chief Technology Officer, Global Patient Identifiers Inc (2012). Voluntary Universal Healthcare Identification System [Online]. Retrieved from http://www.gpii.info/system.php

Hillestad, R., Bigelow, J. H., Chaudhry, B., Dryer, P., Greenberg, M. D., Meili, R. C., Ridgely, M. S., Rothenberg, J., & Taylor, R. (2008). Identify crisis: An examination of the costs and benefits of a unique patient identifier for the U.S. health care system. *Rand Health* [Online]. Available at http://www.rand.ord/pibs/monographs/MG753

John P. Holdren, Co-Chair, Eric Lander, Co-Chair (2010). Report to the President Realizing the Full Potential of Health Information Technology to Improve Healthcare for Americans: The Path Forward, Executive Office of the President, President's Council of Advisors on Science and Technology

Jacques, L. B. (2011). Electronic health records and respect for patient privacy: A prescription for compatibility. *Vanderbilt Journal of Entertainment and Technology Law 13*(2), 441-461. Retrieved from http://www.jetlaw.org/wp-content/journal-pdfs/Francis.pdf

Katish, E., Sondheimer, N., Dullah, P., & Stromberg, S. (2011). Is there an app for that? Electronic health records (ENRS) and a new environment of conflict prevention and resolution. *Law and Contemporary Problems, 74*(31), 31-56. Retrieved from http://scholarship.law.duke.edu/lcp/vol74/iss3/3

Kizer, K. (2007). The adoption of electronic health records: benefits and challenges. *Annals of Health Law, 16*(2), 323-334. Available at http://www.ncbi.nlm.nih.gov/pubmed/17982826

Manos, D. (2012, Feb. 1). Reported health data breaches rose by 97% in 2011, report find. *Healthcare IT News* [Online] Retrieved from http://www.ihealthbeat.org/ articles/2012/2/1/health-data-breaches-increased-by-97-in-2011-report-finds.aspx#ixzz1lKlnpwXP

Miller, S. J. (2008). Electronic medical records: How the potential for misuse outweighs the benefits of transferability. *Journal of Health and Biomedical Law, 4*(2), 353-373.

Niland, Joyce C., Ph.D., Associate Director & Chair, Information Sciences, City of Hope, HIT Policy Committee & HIT Standards Committee, PCAST Workgroup, February 15, 2011, Panel 3, Population Health Written Testimony

Roman, L. (2009). Combined EMR, EHR, and PHR manage data for better health. *Drug Store News* [Online]. Retrieved from http://fmdarticles.com/p/articles/mi_m3374/ is_9_31/ai_n35619257

*Science Daily.* (2008, Oct. 20). Creating unique health ID numbers would improve health care quality, efficiency, study claims [Online]. Retrieved from http://www.rand.org/news/press/2008/10/20.html

Stead, W. W. (2007). Rethinking electronic health records to better achieve quality and safety goal. *Annual Review of Medicine, 36* [Online]. Available at http://arjournals.annualreviewsorg/doi/abs/10.1146/annurev.med.58.061705.144942

Sterstein, A. (2011). Coburn: Computerized records will bring on hackers. *National Journal* [Online]. Retrieved from http://mobile.nationaljournal.com/healthcare/coburn-computerized-patient-records-will-bring-on-hackers-20111110

### Notes to Chapter 15

Bipartisan Policy Center (2012, June) Challenges and Strategies for Accurately Matching Patients to Their Health Data. Retrieved from http://www.redwoodmednet.org/projects/events/20120719/Accuratel

y-Matching-Patients-to-Their-Health-Data.pdf

College of Healthcare Information Management Executives. (2012). Summary of CHIME survey on patient data-matching. Presentation during May 16, 2012 meeting of Bipartisan Policy Center Collaborative on Health IT and Delivery System Reform. Washington, D.C. Retrieved from http://www.redwoodmednet.org/projects/events/20120719/Accuratel y-Matching-Patients-to-Their-Health-Data.pdf

iHealthBeat (2013, December 17) ONC Releases Findings From Patient Data Matching Study. Retrieved from http://www.ihealthbeat.org/articles/2013/12/17/onc-releases-findings-from-patient-data-matching-study

Li, X., & Shen, C. (2013). Linkage of patient records from disparate sources. *Statistical Methods In Medical Research*, *22*(1), 31-38. doi:10.1177/0962280211403600

Paxton, A. (2009). National patient ID: could a voluntary system fill the gap? College of American Pathologists. Retrieved from http://www.cap.org/apps/cap.Portal?_nfpb=true&cntvwrPtlt_actionO verride=%2Fportlets%2FcontentViewer%2Fshow&_windowLabel=cnt vwrPtlt&cntvwrPtlt%7BactionForm.contentReference%7D=cap_today %2F1109%2F1109j_national_id.html&_state=maximiz ed&_pageLabel=cntvwr

Peel, D.C. (2013), December 16). Patient Identification and Matching Initial Findings.Patient Privacy Rights. Retrieved from http://patientprivacyrights.org/wp-content/uploads/2013/12/PPR-Patient-Matching-Testimony-for-12.16.13.pdf

Scheuren F. (1997). Linking health records: human rights concerns.In: Proceedings of an International Workshop and Exposition:Record Linkage Techniques, Arlington, USA

Stead, W.W., Kelly, B.J., & Kolodner, R.M. (2005). Achievable steps toward building a national health information infrastructure in the United States. *Journal of the American Medical Information Association*, 12(2), 113-120

## Index

## About the Author

Tim Godlove, Ph.D, is experienced in multidimensional administrative, information technology, information assurance, project and risk management, and strategic planning. A retired U.S. Navy Master Chief and he has earned a Doctor of Philosophy (Ph.D.) in Information Systems, Information Assurance Policy from the University of Fairfax, MSA in Information Resource Management from Central Michigan University, and his BA from Chapman University. Holds the following FITSP-M, CNSS/NSA 4012/4011 and completed the Chief Information Officer Program and Information Assurance Certification at the National Defense University.

## Contributing Author

Mr. Adrian W. Ball is a PMP and ITIL-certified IT Program Manager with over 30 years of experience organizing, planning and executing custom and commercial off-the-shelf IT programs for commercial and government clients. Mr. Ball has established IT project management offices at both federal government agencies and contracting companies. He's currently the program manager for a $100M five-year HHS/CMS software development contract with TurningPoint Global Solutions, a Rockville, MD based CMMI Level 3 organization. Mr. Ball holds a Master's Degree in Computer Science from Rensselaer Polytechnic Institute and a Bachelor's Degree in Mechanical Engineering from the University of Rhode Island.

www.ingramcontent.com/pod-product-compliance
Lightning Source LLC
Chambersburg PA
CBHW032330210326
41518CB00041B/2000